The IMA Volumes
in Mathematics
and its Applications

Volume 72

Series Editors
Avner Friedman   Willard Miller, Jr.

Institute for Mathematics and
its Applications
IMA

The **Institute for Mathematics and its Applications** was established by a grant from the National Science Foundation to the University of Minnesota in 1982. The IMA seeks to encourage the development and study of fresh mathematical concepts and questions of concern to the other sciences by bringing together mathematicians and scientists from diverse fields in an atmosphere that will stimulate discussion and collaboration.

The IMA Volumes are intended to involve the broader scientific community in this process.

Avner Friedman, Director
Willard Miller, Jr., Associate Director

* * * * * * * * * *

IMA ANNUAL PROGRAMS

1982–1983	Statistical and Continuum Approaches to Phase Transition
1983–1984	Mathematical Models for the Economics of Decentralized Resource Allocation
1984–1985	Continuum Physics and Partial Differential Equations
1985–1986	Stochastic Differential Equations and Their Applications
1986–1987	Scientific Computation
1987–1988	Applied Combinatorics
1988–1989	Nonlinear Waves
1989–1990	Dynamical Systems and Their Applications
1990–1991	Phase Transitions and Free Boundaries
1991–1992	Applied Linear Algebra
1992–1993	Control Theory and its Applications
1993–1994	Emerging Applications of Probability
1994–1995	Waves and Scattering
1995–1996	Mathematical Methods in Material Science

IMA SUMMER PROGRAMS

1987	Robotics
1988	Signal Processing
1989	Robustness, Diagnostics, Computing and Graphics in Statistics
1990	Radar and Sonar (June 18 - June 29)
	New Directions in Time Series Analysis (July 2 - July 27)
1991	Semiconductors
1992	Environmental Studies: Mathematical, Computational, and Statistical Analysis
1993	Modeling, Mesh Generation, and Adaptive Numerical Methods for Partial Differential Equations
1994	Molecular Biology

* * * * * * * * * *

SPRINGER LECTURE NOTES FROM THE IMA:

The Mathematics and Physics of Disordered Media

Editors: Barry Hughes and Barry Ninham
(Lecture Notes in Math., Volume 1035, 1983)

Orienting Polymers

Editor: J.L. Ericksen
(Lecture Notes in Math., Volume 1063, 1984)

New Perspectives in Thermodynamics

Editor: James Serrin
(Springer-Verlag, 1986)

Models of Economic Dynamics

Editor: Hugo Sonnenschein
(Lecture Notes in Econ., Volume 264, 1986)

David Aldous   Persi Diaconis
Joel Spencer   J. Michael Steele
Editors

Discrete Probability
and Algorithms

With 7 Illustrations

Springer-Verlag

New York Berlin Heidelberg London Paris
Tokyo Hong Kong Barcelona Budapest

David Aldous
Department of Statistics
University of California
Berkeley, CA 94720 USA

Persi Diaconis
Department of Mathematics
Harvard University
Cambridge, MA 02138 USA

Joel Spencer
Department of Computer Science
Courant Institute
New York University
251 Mercer Street
New York, NY 10012 USA

J. Michael Steele
Department of Statistics
The Wharton School
University of Pennsylvania
3000 Steinberg Hall-Dietrich Hall
Philadelphia, PA 19104-6302 USA

Mathematics Subject Classifications (1991): 60-06, 05-06, 68-06

Library of Congress Cataloging-in-Publication Data
Discrete probability and algorithms / David Aldous . . . [et al.],
    editors.
        p.   cm. — (The IMA volumes in mathematics and its
    applications ; v. 72)
        Papers from two workshops held at the University of Minnesota in fall
    1993.
        Includes bibliographical references.
        ISBN 0-387-94532-6 (hardcover)
        1. Probabilities—Congresses.   2. Algorithms—Congresses.
    I. Aldous, D. J. (David J.)   II. Series.
    QA273.A1D57   1995
    519.2—dc20   95-151

Printed on acid-free paper.

Production managed by Hal Henglein; manufacturing supervised by Joe Quatela.
Camera-ready copy prepared by the IMA.
Printed and bound by Braun-Brumfield, Ann Arbor, MI.
Printed in the United States of America.

9 8 7 6 5 4 3 2 1

ISBN 0-387-94532-6 Springer-Verlag New York Berlin Heidelberg

The IMA Volumes
in Mathematics and its Applications

Current Volumes:

Volume 67: Mathematics in Industrial Problems, Part 7
  by Avner Friedman

Volume 68: Flow Control
  Editor: Max D. Gunzburger

Volume 69: Linear Algebra for Signal Processing
  Editors: Adam Bojanczyk and George Cybenko

Volume 70: Control and Optimal Design of Distributed Parameter
  Systems
  Editors: John E. Lagnese, David L. Russell, and Luther W. White

Volume 71: Stochastic Networks
  Editors: Frank P. Kelly and Ruth J. Williams

Volume 72: Discrete Probability and Algorithms
  Editors: David Aldous, Persi Diaconis, Joel Spencer, and
  J. Michael Steele

Forthcoming Volumes:

1992 Summer Program: *Environmental Studies: Mathematical,
  Computational, and Statistical Analysis*

1992–1993: *Control Theory*

  Robotics

  Nonsmooth Analysis & Geometric Methods in Deterministic Optimal
  Control

1993 Summer Program: *Modeling, Mesh Generation, and
  Adaptive Numerical Methods for Partial Differential Equations*

1993-1994: *Emerging Applications of Probability*

  Random Discrete Structures

  Mathematical Population Genetics

  Stochastic Problems for Nonlinear Partial Differential Equations

  Image Models (and their Speech Model Cousins)

  Stochastic Models in Geosystems

  Classical and Modern Branching Processes

1994 Summer Program: *Molecular Biology*

1994-1995: *Waves and Scattering*

    Computational Wave Propagation

    Wavelets, Multigrid and Other Fast Algorithms (Multipole, FFT)
      and Their Use In Wave Propagation

    Waves in Random and Other Complex Media

FOREWORD

This IMA Volume in Mathematics and its Applications

DISCRETE PROBABILITY AND ALGORITHMS

is based on the proceedings of two workshops, "Probability and Algorithms" and "The Finite Markov Chain Renaissance" that were an integral part of the 1993–94 IMA program on "Emerging Applications of Probability." We thank David Aldous, Persi Diaconis, Joel Spencer, and J. Michael Steele for organizing these workshops and for editing the proceedings. We also take this opportunity to thank the National Science Foundation, the Air Force Office of Scientific Research, the Army Research Office, and the National Security Agency, whose financial support made the workshop possible.

Avner Friedman

Willard Miller, Jr.

PREFACE

Discrete probability theory and the theory of algorithms have become close partners over the last ten years, though the roots of the partnership go back much longer. There are many reasons that underlie the coordination of these two fields, but some sense of the driving principles can be evoked by considerations like the following:

- When the use of a rule in an algorithm might lead to locking conflicts, randomization often provides a way to avoid stalemate.
- When a combinatorial object cannot be easily constructed, one can still often show the existence of the object by showing that under a suitable probability model such an object (or one close enough for appropriate modification) will exist with positive probability.
- When one needs to make a random uniform selection from an intractably large set, one can sometimes succeed by making clever use of a random walk (or other Markov chain) that has for its stationary measure the desired distribution.
- Finally, in many large systems that are driven by elements of chance, one often finds a certain steadiness that can be expressed by limit laws of probability theory and that can be exploited in the design of algorithms.

All of the chapters in this volume touch on one or more of these themes. The method of probabilistic construction is at the heart of the paper by Spencer and Tetali on Sidon sets as well as that of Godbole, Skipper, and Sunley, which traces its roots back to one of the first great successes of the "probabilistic method" — Erdős's pioneering analysis of the central Ramsey numbers.

The theme of steadiness in large random structures is evident in almost all of the volume's chapters, but it is made explicit in the paper by Fill and Dobrow on the move-to-front rule for self-organizing lists, the chapter by Yukich on Euclidean functionals (like the TSP), and in the paper by Steele that explores the limit theory that has evolved from the Erdős-Szekeres theorem on monotone subsequences. The chapter by Alon also shows how to find algorithmically useful "order in chaos" by developing a basic criterion of network connectivity in random graph models with unequal probabilities for edges.

The theme of "uniform selection by walking around" is perhaps most explicitly illustrated in the two chapters by Diaconis and Gangolli and Diaconis and Holmes. The first of these shows how one can use the ideas of the "Markov Chain Renaissance" to make progress on the difficult problem of the enumeration of integer tables with specified row sums and column sums. The second paper shows how new developments emerging from the Markov chain renaissance can be brought to bear on problems of concern in statistics, computer science, and statistical mechanics. Aldous also makes a contribution in the thick of the new theory of finite Markov chains by showing in his chapter that one can simulate an observation from a chain's stationary distribution (quickly, though approximately)— all the while not knowing the transition probabilities of the chain except through the action of a "take a step from state x" oracle.

The two further chapters in this collection are surveys that call on all of the basic themes recalled above. They are also tightly tied to the central concerns of the theory of probabilistic complexity. The first of these is the survey of A. Karlin and P. Raghavan on random walks in undirected graphs—a notion that is present in many of the collection's chapters. The second is the survey by D. Welsh on randomized approximation schemes for Tutte-Gröthendieck invariants, which are remarkable polynomials whose values at special points give precise information about such basic graph theoretic problems as the number of connected subgraphs, the number of forest subgraphs, the number of acyclic orientations, and much more.

All the papers in this volume come from two Workshops, "Probability and Algorithms" and "The Finite Markov Chain Renaissance," that were held during the Special Year in Emerging Applications of Probability at the Institute for Mathematics and Its Applications at the University of Minnesota during the fall of 1993. The IMA provided a singularly congenial environment for productive scientific exchange, and with any luck the chapters of this volume will convey a sense of the excitement that could be felt in the progress that was reported in these IMA Workshops.

It is a pleasure to thank Avner Friedman, Willard Miller, Jr., and the IMA staff for their efficient organization of the workshops and the entire program, and to thank Patricia V. Brick for administering the preparation of this volume.

David Aldous
Persi Diaconis
Joel Spencer
J. Michael Steele

CONTENTS

ON SIMULATING A MARKOV CHAIN STATIONARY DISTRIBUTION WHEN TRANSITION PROBABILITIES ARE UNKNOWN[*]

DAVID ALDOUS[†]

Abstract. We present an algorithm which, given a n-state Markov chain whose steps can be simulated, outputs a random state whose distribution is within ε of the stationary distribution, using $O(n)$ space and $O(\varepsilon^{-2}\tau)$ time, where τ is a certain "average hitting time" parameter of the chain.

1. Introduction. Our topic is a small corner of the region where algorithms and Markov chains meet. While perhaps not relevant to the main theoretical or practical issues in that region, it seems interesting enough to be worth recording. My motivation came from a remarkable result of Asmussen, Glynn and Thorisson [2], restated as Theorem 1 below.

Consider a Markov chain on states $\{1, 2, \ldots, n\}$ with irreducible transition matrix $\mathbf{P} = (p(i,j))$ and hence with a unique stationary distribution $\pi_{\mathbf{P}} = (\pi_{\mathbf{P}}(i))$. Suppose we have a subroutine that simulates steps from $\mathbf{P}$, i.e. given any state i as input it outputs a random state J_i with $P(J(i) = j) = p(i,j)$ $\forall j$, independent of previous output. The problem is to devise an algorithm which terminates in some random state ξ such that, regardless of $\mathbf{P}$,

$$(1.1) \qquad \frac{1}{2}\sum_j |P(\xi = j) - \pi_{\mathbf{P}}(j)| \le \varepsilon.$$

The point is that the algorithm is not allowed to know $\mathbf{P}$. That is, given the first s steps $(i_1, j_1), (i_2, j_2), \ldots, (i_s, j_s)$, we must specify a rule by which we either terminate and output j_s, or else specify a state i_{s+1} to be the next input to the subroutine, and this rule can use only $(i_1, j_1, i_2, j_2, \ldots, j_s)$ and external randomization. Write $\mathcal{A}(\varepsilon)$ for the class of algorithms A which satisfy (1.1) for all irreducible $\mathbf{P}$ (we suppress dependence on n here). For $A \in \mathcal{A}(\varepsilon)$ let $c(A, \mathbf{P})$ be the mean number of steps simulated by the algorithm.

It is obvious that (even for $n = 2$) there is no algorithm $A \in \mathcal{A}(\varepsilon)$ such that $\sup_{\mathbf{P}} c(A, \mathbf{P}) < \infty$, by considering "almost reducible" chains.

Two different methods for attempting to construct algorithms in $\mathcal{A}(\varepsilon)$ suggest themselves.

Matrix perturbation theory gives bounds for $\pi_{\mathbf{Q}} - \pi_{\mathbf{P}}$ is terms of bounds for $\mathbf{Q} - \mathbf{P}$. Fix some integer m, simulate m steps from each state i to get an empirical estimate

* Research supported by N.S.F. Grant DMS92-24857.

† Department of Statistics, University of California, Berkeley, CA 94720.

$Q_m(i, j)$ of the $p(i, j)$ together with an estimate of the error $\mathbf{Q}_m - \mathbf{P}$. Then calculate numerically the stationary distribution π_m corresponding to $\mathbf{Q}_m$ and a confidence interval for the error $\pi_m - \pi$. If this interval is too large, increment m and repeat.

Let's call this a *matrix-based* method, in contrast to a *pure simulation* method below.

Markov chain theory says that if we simulate the chain from an arbitrary initial state then at a sufficiently large time t (randomized to avoid periodicity) the current state will have approximately distribution π. So fix t, simulate the chain for t steps, perform some test on the observed path to check if t is sufficiently large, and if not then increment t and repeat.

It is of course not entirely clear how to turn these vague ideas into algorithms which are provably in $\mathcal{A}(\varepsilon)$. Using a rather different idea, Asmussen et al ([2] Theorem 3.1) showed that in fact one can simulate $\pi_\mathbf{P}$ exactly with no knowledge of $\mathbf{P}$.

THEOREM 1. *There exists an $A_0 \in \mathcal{A}(0)$ with $c(A_0, \mathbf{P}) < \infty \ \forall \mathbf{P}$.*

Briefly, given a way of simulating exactly the distribution π restricted to $B_j = \{1, 2, \ldots, j\}$, then they describe a procedure to simulate exactly the distribution π restricted to B_{j+1}.

Unfortunately it seems difficult to give an informative upper bound for $c(A_0, \mathbf{P})$ in terms of $\mathbf{P}$. In section 2 we present and analyze a slightly simpler "pure simulation" algorithm which does permit a natural upper bound, at the cost of producing approximate rather than exact stationarity. Theorem 2 states the precise result. Section 3 gives a lower bound for the performance of any algorithm, and section 4 contains further discussion.

2. The algorithm. We start by outlining the idea of the algorithm. Write $E_i T_j$ for the mean hitting time on state j, starting from state i. Define the averaged hitting time $\tau = \tau_\mathbf{P}$ by

$$(2.1) \qquad\qquad \tau = \sum_i \sum_j \pi(i) E_i T_j \pi(j).$$

Because τ is essentially an upper bound on the time taken to approach stationarity, it is enough to be able to estimate the value of τ by simulation in $O(\tau)$ steps, for then we can run another simulation for $O(\tau)$ steps and output the final state. The estimation of τ is done via a "coalescing paths" routine: run the chain from an arbitrary start until a specified state j is hit, keeping track of states visited; start again from some unvisited state and run until visiting some state hit on a previous run; and so on until every state has been visited. Then the number of steps used in this procedure is $\Theta(\tau)$, for a typical initial target j.

Here is the precise algorithm. We are given $\varepsilon > 0$, states $\{1, \ldots, n\}$, and the ability to simulate a step of the Markov chain from any specified state.

Algorithm A_ε
1. Let $t_0 \leftarrow n$.
2. Pick U random, uniform on $\{1, \ldots, t_0\}$.
3. Simulate the chain, starting at state 1, for U steps. Let j be the final state.
4. Start a counter at 0 and count steps as they are simulated in the stages below. If the count exceeds εt_0 before this algorithm starts stage **8**, let $t_0 \leftarrow 2t_0$ and go to stage **2**.
5. Let $B \leftarrow \{j\}$.
6. Simulate the chain, starting at state $\min\{i : i \notin B\}$, until the chain hits B, keeping track of the set B' of states visited.
7. Let $B \leftarrow B \cup B'$. If $B \neq \{1, \ldots, n\}$ go to stage **6**.
8. Pick U random, uniform on $\{1, \ldots, \lceil t_0/\varepsilon \rceil\}$. Simulate the chain, starting at state 1, for U steps. Output the final state ξ.

Stages **5,6,7** are the "coalescing walks" routine. It is clear that the algorithm requires only $O(n)$ space, to track which states are hit during this routine.

THEOREM 2. *Fix $0 < \varepsilon < 1/4$. Then $A_\varepsilon \in \mathcal{A}(4\varepsilon)$ and*

$$c(A_\varepsilon, \mathbf{P}) \leq \frac{81\tau_{\mathbf{P}}}{\varepsilon^2} \ \forall \mathbf{P}.$$

One might guess that some variation of the construction would lead to an algorithm where the bound is polynomial in $\log 1/\varepsilon$, but I have not pursued that possibility.

The rest of the section contains the proof of Theorem 2. To start with some notation, write $\| \ \|$ for variation distance between distributions

$$\|\theta - \mu\| = \frac{1}{2} \sum_i |\theta(i) - \mu(i)|.$$

Write

$$s(j) = \max_i E_i T_j$$

$$s_* = \min_j s(j).$$

Write C_j for the "coalescing paths" time, that is the number of steps required to complete stages **5,6,7** of the algorithm, ignoring the cut-off rule in **4**. Finally, recall two standard facts: the *right-averaging principle* ([1] Chapter 2)

(2.2) $$\sum_j E_i T_j \ \pi(j) = \tau \ \forall i$$

and a renewal identity ([1] Chapter 2)

(2.3) $E_i($ number of visits to i before $T_j) = \pi(i)(E_iT_j + E_jT_i)$, $j \neq i$

where "number of visits" includes the visit at time 0.

LEMMA 3. *For states $j \neq i$ and times $u \geq 0$,*

$$P(C_j \leq u) \leq P_i(T_j \leq u).$$

Proof. It is enough to exhibit, in the context of the coalescing walks routine, a r.v. $\hat{T} \leq C_j$ such that $\hat{T}$ is distributed as the first hitting time on j started at i. But this can be done in the obvious way. When the coalescing walks routine first hit i there was some current target subset B_1, and the walk followed a trajectory $i = i_0, i_1, \ldots, i_{y_1} \in B$. If $i_{y_1} \neq j$ then we proceed recursively; when the coalescing walks routine first hit i_{y_1} there was some current target subset B_2, and the walk followed a trajectory $i_{y_1}, i_{y_1+1}, \ldots, i_{y_2} \in B$. Continue until j is reached, and concatenate these trajectories. This gives a walk from i to j of length $\hat{T} \leq C_j$, and it is straightforward to verify the walk is distributed as the Markov chain started at i and run until first hitting j.

LEMMA 4. $P(C_j \leq u) \leq u/s(j)$, $u > 0$, $\forall j$.

Proof. Fix j and u. Write $b = \max_i P_i(T_j > u)$. By iteration,

$$s(j) \equiv \max_i E_iT_j \leq \frac{u}{1-b}.$$

Rearranging,

$$\min_i P_i(T_j \leq u) \leq u/s(j).$$

So the result follows from Lemma 3.

LEMMA 5. *Let U be uniform on $\{1, 2, \ldots, t\}$, independent of the chain. Then*

$$||P_i(X_U \in \cdot) - \pi(\cdot)|| \leq \frac{\min(\tau, 2s_*)}{t} \quad \forall i.$$

Proof. Let K be a random state with distribution π. Then

$$
\begin{aligned}
(2.4) \quad ||P_i(X_U \in \cdot) - \pi(\cdot)|| &= ||P_i(X_U \in \cdot) - P_i(X_{T_K+U} \in \cdot)|| \\
&\leq ||\text{dist}(U) - \text{dist}(U + T_K)|| \\
&= E_i \min(1, T_K/t) \\
(2.5) \quad &\leq E_i T_K/t \\
&= \tau/t \text{ by (2.2).}
\end{aligned}
$$

Now let j attain the *min* in the definition of s_*. Then

$$
\begin{aligned}
||P_i(X_U \in \cdot) - P_j(X_U \in \cdot)|| &= ||P_i(X_U \in \cdot) - P_i(X_{T_j+U} \in \cdot)|| \\
&\leq E_iT_j/t \text{ repeating the argument (2.4)–(2.5) above} \\
&\leq s_*/t.
\end{aligned}
$$

Averaging over i gives

$$||\pi(\cdot) - P_j(X_U \in \cdot)|| \leq s_*/t$$

and so the triangle inequality gives the bound $2s_*/t$ for the quantity in the Lemma.

LEMMA 6. $EC_j \leq \tau + E_\pi T_j \ \forall j$.

Proof. $EC_j = \sum_{i \neq j} v_i$, where v_i is the expected number of steps from i simulated during the "coalescing walks" routine. And for fixed i,

$$v_i = E_i(\text{ number of visits to } i \text{ before } T_B)$$

for some set $B \ni j$

$$\leq E_i(\text{ number of visits to } i \text{ before } T_j)$$

$$= \pi(i)(E_i T_j + E_j T_i) \text{ by (2.3).}$$

Summing over i,

$$EC_j \leq E_\pi T_j + \sum_i E_j T_i \ \pi(i)$$

and the result follows from (2.2).

LEMMA 7. *Let* $J = X_U$, *where* $X_0 = 1$ *and* U *is uniform on* $\{1, 2, \ldots, t\}$ *independent of the chain. Use independent realizations to construct* C_J. *Then*

$$P(C_J \geq \varepsilon t) \leq 3\sqrt{\frac{\tau}{\varepsilon t}}.$$

Proof. By Markov's inequality and Lemma 6,

(2.6) $$P(C_j \geq \varepsilon t) \leq \frac{EC_j}{\varepsilon t} \leq \frac{\tau + E_\pi T_j}{\varepsilon t}.$$

Write $t(j) = E_\pi T_j$. By Markov's inequality, for any $x > 0$ we have $\pi\{j : t(j) > x\} \leq \tau/x$, and then by Lemma 5

$$P(t(J) > x) \leq \frac{\tau}{x} + \frac{\tau}{t}.$$

Substituting into (2.6),

$$P(C_J \geq \varepsilon t) \leq \frac{\tau + x}{\varepsilon t} + \frac{\tau}{t} + \frac{\tau}{x}.$$

Putting $x = \sqrt{\varepsilon t \tau}$ the bound becomes

$$\frac{\tau}{t}\left(1 + \frac{1}{\varepsilon}\right) + 2\sqrt{\frac{\tau}{\varepsilon t}}.$$

We may assume $\sqrt{\frac{\tau}{\varepsilon t}} \leq 1/3$, and then it is easy to check $\frac{\tau}{t}(1+\frac{1}{\varepsilon}) \leq \sqrt{\frac{\tau}{\varepsilon t}} \leq 1/3$, and the Lemma follows.

Proof of Theorem 2. The algorithm enters the "coalescing walks" routine with successive values $t_0 = n, 2n, 2^2 n, 2^3 n, \ldots$ and fails to complete the routine within the required εt_0 steps until finally succeeding with some $t_0 = n2^M$, say. Granted it enters with $t_0 = n2^m$, Lemma 4 implies that the chance it succeeds with that t_0 is at most $\varepsilon n 2^m / s_*$. So the chance the algorithm terminates with some $t_0 \leq s_*$ is at most 2ε. Now suppose the algorithm enters stage **8** with some $t_0 \geq s_*$. Then by Lemma 5

$$\|P(\xi \in \cdot) - \pi(\cdot)\| \leq \frac{2s_*}{\lceil t_0/\varepsilon \rceil} \leq 2\varepsilon.$$

So overall the variation distance between dist(ξ) and π is $\leq 4\varepsilon$. That is, $A_\varepsilon \in \mathcal{A}(4\varepsilon)$.

To study the number of steps of the chain simulated by the algorithm, suppose it terminates with $t_0 = n2^M$. Then the number of steps is at most

$$(2.7) \qquad n2^{M+1}(1 + \varepsilon) + n2^M/\varepsilon \leq n2^{M+1}/\varepsilon$$

where the final inequality uses the assumption $\varepsilon < 1/4$. We now bound M. Choose m_0 to be the smallest integer such that

$$\beta \equiv 3\sqrt{\frac{\tau}{\varepsilon n 2^{m_0}}} \leq 1.$$

Lemma 7 implies that, if the algorithm enters the coalescing walks routine with $t_0 = n2^{m_0+m}$, the chance that it fails to complete the routine before the cut-off is at most $\beta 2^{-m/2} \leq 2^{-m/2}$. Thus

$$P(M - m_0 > m) \leq \prod_{u=1}^{m} 2^{-u/2}$$

and a brief calculation shows this implies $E2^{M-m_0} \leq 9/4$. Substituting into (2.7), the expected number of steps is at most $\frac{9n2^{m_0}}{2\varepsilon}$. But from the definition of m_0 we have $n2^{m_0} \leq \frac{18\tau}{\varepsilon}$ and the Theorem follows.

3. A lower bound lemma. As a prelude to the discussion in section 4 we give a lower bound on the mean number of simulation steps required by any algorithm. The statement looks complicated, but the reader will see that it's just what comes out of "the obvious argument".

Write

$$c_*(\mathbf{P}, \varepsilon) = \inf_{A \in \mathcal{A}(\varepsilon)} c(A, \mathbf{P}).$$

LEMMA 8.

$$c_*(\mathbf{P}, \varepsilon) \geq \inf_{\theta} \sup_{\mathbf{Q}} \frac{\|\pi_{\mathbf{Q}} - \pi_{\mathbf{P}}\| - 2\varepsilon}{\frac{1}{2}\sum_i \sum_j \theta(i)|q(i,j) - p(i,j)|}$$

where θ denotes an irreducible probability distribution and $\mathbf{Q}$ a transition matrix on $\{1, 2, \ldots, n\}$.

Proof. Fix $\mathbf{P}, \varepsilon$ and $A \in \mathcal{A}(\varepsilon)$. Let $m(i)$ be the mean number of steps from state i simulated when the algorithm A is used on transition matrix $\mathbf{P}$. For any $\mathbf{Q}$ there is a natural coupling of the $\mathbf{P}$-chain and the $\mathbf{Q}$-chain under which the chance that the steps from i in the two chains go to different states is $d(i) = \frac{1}{2} \sum_j |p(i,j) - q(i,j)|$. By considering coupled realizations of the chains it is easy to see that the output states $\xi^{\mathbf{P}}$ and $\xi^{\mathbf{Q}}$ satisfy

$$P(\xi^{\mathbf{Q}} \neq \xi^{\mathbf{P}}) \leq \sum_i d(i)m(i).$$

So

$$||P(\xi^{\mathbf{P}} \in \cdot) - P(\xi^{\mathbf{Q}} \in \cdot)|| \leq \sum_i d(i)m(i).$$

But the fact $A \in \mathcal{A}(\varepsilon)$ implies $||P(\xi^{\mathbf{Q}} \in \cdot) - \pi_{\mathbf{Q}}|| \leq \varepsilon$ and the analog for $\mathbf{P}$, so

$$||\pi_{\mathbf{P}} - \pi_{\mathbf{Q}}|| - 2\varepsilon \leq \sum_i d(i)m(i).$$

The measure $m(\cdot)$ is just $c(A, \mathbf{P})$ times some probability distribution θ, so rearranging gives

$$c(A, \mathbf{P}) \geq \frac{||\pi_{\mathbf{Q}} - \pi_{\mathbf{P}}|| - 2\varepsilon}{\frac{1}{2} \sum_i \sum_j \theta(i)|q(i,j) - p(i,j)|}.$$

This holds for all $\mathbf{Q}$, so the Lemma follows.

4. Discussion. The problem considered in this paper is intrinsically harder for some $\mathbf{P}$'s than for others. This is the context in which it is reasonable to compare algorithms, not by worst-case or average-case analysis, but by *competitive analysis* (see e.g. [4]). That is, we seek an algorithm $A \in \mathcal{A}(\varepsilon)$ which makes the *competitive ratio*

$$\text{comp}(A, \varepsilon) \equiv \sup_{\mathbf{P}} \frac{c(A, \mathbf{P})}{c_*(\mathbf{P}, \varepsilon)} \geq 1$$

as small as possible. I conjecture that, in the setting where the cost $c(A, \mathbf{P})$ involves only the number of steps of the chain simulated, and where matrix computations are free, there is some "matrix-based" algorithm whose competitive ratio is polynomial in $\log n$ and $1/\varepsilon$. To be more realistic one should assign costs to matrix computations also. It would be an interesting research project to carry through a competitive analysis in such a setting.

One can ask whether, in the case where $\mathbf{P}$ is known, it is ever better to use an algorithm like ours to generate a realization from the stationary distribution, rather than matrix computations. Certainly there might be ranges of n for which general-purpose matrix packages might not run on one's workstation, and where it would be easier to code a simulation rather than write special purpose matrix code. In particular, if one is intending to do a simulation study of the stationary chain, so the issue is to choose an initial state, and if a heuristic guess at τ is a practical number of steps to simulate, then there is nothing to lose by trying the algorithm. The workshop proceedings [5] give an overview of numerical methods in Markov chains.

It is easy to see that $\tau = \Omega(n)$. The heuristic interpretation of the ratio τ/n is as the expectation, starting at a typical state i, of the number of returns to i before approaching the stationary distribution. Thus the property $\tau = \Theta(n)$ is a kind of "local transience" property. For the (discretized) $M/M/1/n$ queue (states $\{0, 1, \ldots, n-1\}$ and transitions $i \to i \pm 1$ with $p(i, i+1) = p < 1/2$) we have $\tau \sim \frac{1}{1-2p} n$, and I believe that most classical stable queueing networks should have $\tau = \Theta(n)$. This will fail for networks exhibiting bistability, e.g. circuit-switched networks under high load (see Kelly [3]).

Note added in proof. A recent preprint by Lovasz and Winkler "Exact Mixing in an Unknown Markov Chain" studies a new algorithm which simulates π exactly. The algorithm is based upon the representation of stationary distributions in terms of weighted spanning trees. Their upper bound for the number of steps is polynomial in $\max_{i,j} E_i T_j$, which is essentially larger than the bound in Theorem 2.

REFERENCES

[1] D.J. Aldous. Reversible Markov chains and random walks on graphs. Book in preparation, 1995.

[2] S. Asmussen, P.W. Glynn, and H. Thorisson. Stationarity detection in the initial transient problem. *ACM Trans. Modeling and Computer Sim.*, 2:130–157, 1992.

[3] F.P. Kelly. Stochastic models of computer communication networks. *J. Royal Statist. Soc.*, B47:379–395, 1985.

[4] L.A. McGeoch and D. Sleator. *Workshop on On-Line Algorithms*. American Math. Soc., Providence, RI, 1992.

[5] W.J. Stewart. *Numerical Solutions of Markov Chains*. Marcel Dekker, New York, 1991.

A NOTE ON NETWORK RELIABILITY*

NOGA ALON[†]

Let $G = (V, E)$ be a loopless undirected multigraph on n vertices, with a probability p_e, $0 \leq p_e \leq 1$ assigned to every edge $e \in E$. Let G_p be the random subgraph of G obtained by deleting each edge e of G, randomly and independently, with probability $q_e = 1 - p_e$. For any nontrivial subset $S \subset V$ let $(S, \overline{S})$ denote, as usual, the cut determined by S, i.e., the set of all edges of G with an end in S and an end in its complement $\overline{S}$. Define $P(S) = \sum_{e \in (S, \overline{S})} p_e$, and observe that $P(S)$ is simply the expected number of edges of G_p that lie in the cut $(S, \overline{S})$. In this note we prove the following.

THEOREM 1. *For every positive constant b there exists a constant $c = c(b) > 0$ so that if $P(S) \geq c \log n$ for every nontrivial $S \subset V$, then the probability that G_p is disconnected is at most n^{-b}.*

The assertion of this theorem (in an equivalent form) was conjectured by Dimitris Bertsimas, who was motivated by the study of a class of approximation graph algorithms based on a randomized rounding technique of solutions of appropriately formulated linear programming relaxations. Observe that the theorem is sharp, up to the multiplicative factor c, by the well known results on the connectivity of the random graph (see, e.g., [2]). In case our G above is simply the complete graph on n vertices, and $p_e = p$ for every edge e, these known results assert that the subgraph G_p, which in this case is simply the random graph $G_{n,p}$, is almost surely disconnected if $p = (1 - \epsilon) \log n / n$, although in this case $P(S) = \Omega(\log n)$ for all S. Theorem 1 can thus be viewed as a generalization to the case of non-uniform edge probabilities of the known fact that if $p > (1 + \epsilon) \log n / n$ then the random graph $G_{n,p}$ is almost surely connected. It would be interesting to extend some other similar known results in the study of random graphs to the non-uniform case and obtain analogous results for the existence of a Hamilton cycle, a perfect matching or a k-factor.

The above theorem is obviously a statement on network reliability. Suppose G represents a network that can perform iff it is connected. If the edges represent links and the failure probability of the link e is q_e, then the probability that G_p remains connected is simply the probability that the network can still perform. The network is reliable if this probability is close to 1. Thus, the theorem above supplies a sufficient condition for a network to be reliable, and this condition is nearly tight in several cases.

Proof of Theorem 1. Let $G = (V, E)$ be a loopless multigraph and suppose

---

* This research was supported in part by the Institute for Mathematics and its Applications with funds provided by the NSF and by the Sloan Foundation, Grant No. 93-6-6.

† Institute for Advanced Study, Princeton, NJ 08540 and Department of Mathematics, Tel Aviv University, Tel Aviv, Israel.

that $P(S) \geq c \log n$ for every nontrivial $S \subset V$. It is convenient to replace G by a graph G' obtained from G by replacing each edge e by $k = c \log n$ parallel copies with the same endpoints and by associating each copy e' of e with a probability $p'_{e'} = p_e/k$. For every nontrivial $S \subset V$, the quantity $P'(S)$ defined by $P'(S) = \sum_{e' \in (S, \bar{S})} p'_{e'}$ clearly satisfies $P'(S) = P(S)$. Moreover, for every edge e of G, the probability that no copy e' of e survives in $G'_{p'}$ is precisely $(1 - p_e/k)^k \geq 1 - p_e$ and hence G_p is more likely to be connected than $G'_{p'}$. It therefore suffices to prove that $G'_{p'}$ is connected with probability at least $1 - n^{-b}$. The reason for considering G' instead of G is that in G' the edges are naturally partitioned into k classes, each class consisting of a single copy of every edge of G. Our proof proceeds in phases, starting with the trivial spanning subgraph of G' that has no edges. In each phase we randomly pick some of the edges of G' that belong to a fresh class which has not been considered before, with the appropriate probability. We will show that with high probability the number of connected components of the subgraph of G' constructed in this manner decreases by a constant factor in many phases until it becomes 1, thus forming a connected subgraph. We need the following simple lemma.

LEMMA 1. *Let $H = (U, F)$ be an arbitrary loopless multigraph with a probability w_f assigned to each of its edges f, and suppose that for every vertex u of H, $\sum_{v \in U, uv \in E} w_{uv} \geq 1$. Let H_w be the random subgraph of H obtained by deleting every edge f of H, randomly and independently, with probability $1 - w_f$. Then, if $|U| > 1$, with probability at least $1/2$ the number of connected components of H_w is at most $(1/2 + 1/e)|U| < 0.9|U|$.*

Proof. Fix a vertex u of H. The probability that u is an isolated vertex of H_w is precisely

$$\prod_{v \in U, uv \in E} (1 - w_{uv}) \leq exp\{-\sum_{v \in U, uv \in E} w_{uv}\} \leq 1/e.$$

By linearity of expectation, the expected number of isolated vertices of H_w does not exceed $|U|/e$, and hence with probability at least $1/2$ it is at most $2|U|/e$. But in this case the number of connected components of H_w is at most

$$2|U|/e + \frac{1}{2}(|U| - 2|U|/e) = (1/2 + 1/e)|U|,$$

as needed. $\square$

Returning to our graph G and the associated graph G', let $E_1 \cup E_2 \cdots \cup E_k$ denote the set of all edges of G', where each set E_i consists of a single copy of each edge of G. For $0 \leq i \leq k$, define G'_i as follows. G'_0 is the subgraph of G' that has no edges, and for all $i \geq 1$, G'_i is the random subgraph of G' obtained from G'_{i-1} by adding to it each edge $e' \in E_i$ randomly and independently, with probability $p'_{e'}$. Let C_i denote the number of connected components of G'_i. Note that as G'_0 has no edges $C_0 = n$ and note that G'_k is simply $G'_{p'}$. Let us call the index i, $(1 \leq i \leq k)$, *successful* if either G'_{i-1} is connected (i.e., $C_{i-1} = 1$) or if $C_i < 0.9C_{i-1}$.

Claim: For every index i, $1 \leq i \leq k$, the conditional probability that i is successful given any information on the previous random choices made in the definition of G'_{i-1} is at least $1/2$.

Proof: If G_{i-1} is connected then i is successful and there is nothing to prove. Otherwise, let $H = (U, F)$ be the graph obtained from G'_{i-1} by adding to it all the edges in E_i and by contracting every connected component of G'_{i-1} to a single vertex. Note that since $P'(S) \geq c \log n = k$ for every nontrivial S it follows that for every connected component D of G'_{i-1}, the sum of probabilities associated to edges $e \in E_i$ that connect vertices of D to vertices outside D is at least 1. Therefore, the graph H satisfies the assumptions of Lemma 1 and the conclusion of this lemma implies the assertion of the claim. □

Observe, now, that if $C_k > 1$ then the total number of successes is strictly less than $- \log n / \log 0.9$ ($< 10 \log_e n$). However, by the above claim, the probability of this event is at most the probability that a Binomial random variable with parameters k and 0.5 will attain a value of at most $r = 10 \log_e n$. (The crucial observation here is that this is the case despite the fact that the events "i is successful" for different values of i are *not* independent, since the claim above places a lower bound on the probability of success given any previous history.) Therefore, by the standard estimates for Binomial distributions (c.f., e.g., [1], Appendix A, Theorem A.1), it follows that if $k = c \log n = (20+t) \log_e n$ then the probability that $C_k > 1$ (i.e., that $G'_{p'}$ is disconnected) is at most $n^{-t^2/2c}$, completing the proof of the theorem. □

Remarks

1. The assertion of Lemma 1 can be strengthened and in fact one can show that there are two positive constants c_1 and c_2 so that under the assumptions of the lemma the number of connected components of the random subgraph H_w is at most $(1 - c_1)|U|$ with probability at least $1 - e^{-c_2|U|}$. This can be done by combining the Chernoff bounds with the following simple lemma, whose proof is omitted

LEMMA 2. *Let $H = (U, F)$ be an arbitrary loopless multigraph with a non-negative weight w_e associated to each of its edges e. Then there is a partition of $U = U_1 \cup U_2$ into two disjoint subsets so that for $i = 1, 2$ and for every vertex $u \in U_i$,*

$$\sum_{uv \in E, \; v \in U_{3-i}} w_{uv} \geq \frac{1}{2} \sum_{uv \in E, \; v \in U} w_{uv}.$$

For our purposes here the weaker assertion of Lemma 1 suffices.

1. It is interesting to note that several natural analogs of Theorem 1 for other graph properties besides connectivity are false. For

example, it is not difficult to give an example of a graph $G = (V, E)$ and a probability function p, together with two distinguished vertices s and t, so that $P(S) \geq \Omega(n/\log n)$ ($\gg \Omega(\log n)$) for all cuts S separating s and t and yet in the random subgraph G_p almost surely s and t lie in different connected components. A simple example showing this is the graph G consisting of $n/10 \log n$ internally vertex disjoint paths of length $10 \log n$ each between s and t, in which $p_e = 1/2$ for every edge e.

2. Another, more interesting example showing that a natural analog of Theorem 1 for bipartite matching fails is the following. Let A and B be two disjoint vertex classes of cardinality n each. Let A_1 be a subset of $c_1 n$ vertices of A and let B_1 be a subset of $c_1 n$ vertices of B, where, say, $1/8 < c_1 < 1/4$. Let H_1 be the bipartite graph on the classes of vertices A and B in which every vertex of A_1 is connected to every vertex of B and every vertex of B_1 is connected to every vertex of A. Let H_2 be a bipartite constant-degree expander on the classes of vertices A and B; for example, a C_2-regular graph so that between any two subsets X of A and Y of B containing at least $c_1 n/2$ vertices each there are at least $c_1 n$ edges (it is easy to show that such a graph exists using a probabilistic construction, or some of the known constructions of explicit expanders). Finally, let $H = (V, E)$ be the bipartite graph on the classes of vertices A and B whose edges are all edges of H_1 or H_2. Define, also, $p_e = 1/(4C_2)$ for every edge e of H. It is not too difficult to check that the following two assertions hold.

(i) There exists a constant $C = C(c_1, C_2) > 0$ so that for every $A' \subset A$ and $B' \subset B$ that satisfy $|A'| + |B'| > n$:

$$\sum_{uv \in E, u \in A', v \in B'} p_{uv} \geq Cn.$$

(ii) The random subgraph H_p of H almost surely does not contain a perfect matching.

The validity of (i) can be checked directly; (ii) follows from the fact that with high probability not many more than $n/4$ edges of H_2 will survive in H_p and the edges of H_1 cannot contribute more than $2c_1 n < n/2$ edges to any matching.

Acknowledgement I would like to thank Dimitris Bertsimas for bringing the problem addressed here to my attention, Joel Spencer for helpful comments and Svante Janson for fruitful and illuminating discussions.

REFERENCES

[1] N. ALON AND J. H. SPENCER, *The Probabilistic Method*, Wiley, 1992.
[2] B. BOLLOBÁS, *Random Graphs*, Academic Press, 1985.

RECTANGULAR ARRAYS WITH FIXED MARGINS

PERSI DIACONIS* AND ANIL GANGOLLI[t]

Abstract. In a variety of combinatorial and statistical applications, one needs to know the number of rectangular arrays of nonnegative integers with given row and column sums. The combinatorial problems include counting magic squares, enumerating permutations by descent patterns and a variety of problems in representation theory. The statistical problems involve goodness of fit tests for contingency tables. We review these problems along with the available techniques for exact and approximate solution.

1. Introduction. Let $\mathbf{r} = (r_1, \cdots, r_m)$ and $\mathbf{c} = (c_1, \cdots, c_n)$ denote positive integer partitions of N. Let $\Sigma_{\mathbf{rc}}$ denote the set of all $m \times n$ nonnegative integer matrices in which row i has sum r_i and column j has sum c_j. Thus $\Sigma_{ij} T_{ij} = N$ for every $T \in \Sigma_{\mathbf{rc}}$. Throughout, we assume $m, n > 1$; otherwise $\Sigma_{\mathbf{rc}}$ has only one element. As will emerge, $\Sigma_{\mathbf{rc}}$ is always nonempty.

When $m = n$ and $r_i \equiv c_j \equiv r$, $\Sigma_{\mathbf{rc}}$ becomes the set of magical squares. The classical literature on these squares is reviewed in Section 2. For general $\mathbf{r}, \mathbf{c}$, the set $\Sigma_{\mathbf{rc}}$ arises in permutation enumeration problems. These include enumerating permutations by descents, enumerating double cosets, and describing tensor product decompositions. Section 3 describes these problems. Sections 4, 5 describe closely related problems in symmetric function theory.

In statistical applications, $\Sigma_{\mathbf{rc}}$ is called the set of contingency tables with given margins $\mathbf{r}$ and $\mathbf{c}$. Tests for independence and more general statistical models are classically quantified by the chi-square distribution. More accurate approximations require knowledge of $\Sigma_{\mathbf{rc}}$. These topics are covered in Section 6.

Remaining sections describe algorithms and theory for enumerating and approximating $|\Sigma_{\mathbf{rc}}|$. Section 7 describes asymptotic approximations. Section 8 describes algorithms for exact enumeration. Section 9 gives complexity results. Section 10 describes Monte Carlo Markov chain techniques. The different sections are independent and may seem to be quite disparate; the link throughout is the set of tables.

Each of these topics has seen active development in recent years. A useful review of the earlier work is in Good and Crook (1977).

2. Magical squares. Let $H_n(r)$ denote the number of $n \times n$ matrices of nonnegative integers each of whose row and column sums is r. Such matrices are called magical squares and were first studied by MacMahon (1916) who showed $H_3(r) = \binom{r+2}{2} + 3\binom{r+3}{4}$. Stanley (1973, 1986) gives a careful review of the history. He has proved that for $r \geq 0$ $H_n(r)$ is a

---

* Dept. of Mathematics, Harvard University, Cambridge, MA 02138.
[t] Silicon Graphics, 2011 N. Shoreline Blvd., Mount View, CA 94043-1389.

polynomial in r of degree exactly $(n-1)^2$. This polynomial is known for $n \leq 6$; see Jackson and Van Rees (1975).

A polynomial of degree $(n-1)^2$ is determined by its values on $(n-1)^2+1$ points. Stanley shows that the polynomial H_n satisfies

$$H_n(-1) = H_n(-2) = \cdots = H_n(-n+1) = 0;$$

$$H_n(-n-r) = (-1)^{n-1}H_n(r), \quad \text{for all} \quad r.$$

Thus, the values $H_n(i)$ for $1 \leq i \leq \binom{n-1}{2}$ determine $H_n(r)$ for all r. The techniques developed below should enable computation of H_7, and perhaps H_8.

The leading coefficient of $H_n(r)$ is the volume of the polytope of $n \times n$ doubly stochastic matrices (Stanley (1986)). This does not have a closed form expression at this writing. Perhaps inspection of the data will permit a reasonable guess. Bona (1994) gives bounds on the volume of the doubly stochastic matrices.

The problem of enumerating magical squares with both diagonals summing to r is discussed in Section 8.2 below. The theory of the present section should apply here but little previous work has been done. We further mention recent work of Jia (1994) which uses multivariate spline techniques to give new proofs of Stanley's results as well as resolve some open problems. Finally, we mention work by Gessel (1990) and Goulden, Jackson, Reilly (1983). These authors show that the problems of this section are P-recursive.

3. Examples in the permutation group. Let S_n be the group of permutations of n objects. This section shows how $\Sigma_{\mathbf{rc}}$ arises in enumerating permutations by descents, in describing double cosets, and in decomposing induced representations and tensor products.

3.1. Descents. A permutation has a descent at i if $\pi(i) > \pi(i+1)$. The descent set of π is $D(\pi) = \{i : \pi(i) > \pi(i+1)\}$. Thus 2431 has descents at positions 2 and 3 so $D(2431) = \{2,3\}$. By definition, $D(\pi) \subseteq \{1,2,\cdots,n-1\}$. There is a useful recoding of descent sets as compositions of $n : D \longmapsto C(D)$. If $D = \{d_1, d_2, \cdots d_r\}$ with $d_1 < d_2 < \cdots < d_r$, set $c_1 = d_1$, $c_2 = d_2 - d_1 \cdots c_{r+1} = n - d_r$. Thus $D = \{2,3\}$ has $C(D) = (2,1,1)$. The map back has $D(C) = \{c_1, c_1 + c_2, \cdots, c_1 + \cdots + c_{r-1}\}$. Descents are actively studied in several areas of mathematics. See, e.g., Stanley (1986), Gessel and Reutenauer (1994), Diaconis, McGrath, Pitman (1993) and the literature cited there. The following result is attributed to Foulkes in the folklore of combinatorics. The elegant bijective proof given below is due to Nantel Bergeron (personal communication).

THEOREM 3.1 (FOULKES). *Let* $\mathbf{r}$ *and* $\mathbf{c}$ *be compositions of* N. *The number of permutations* π *in* S_N *with* $D(\pi) \subseteq D(\mathbf{r})$ *and* $D(\pi^{-1}) \subseteq D(\mathbf{c})$ *is* $|\Sigma_{\mathbf{rc}}|$.

Example 3.2. The following display lists $\pi/D(\pi); \pi^{-1}/D(\pi^{-1})$ for π in S_4

1234/ϕ ; 1234/ϕ	2134/1 ; 2134/1	3124/1 ; 2314/2	4123/1 ; 2341/3
1243/3 ; 1243/3	2143/13 ; 2143/13	3142/13 ; 2413/2	4132/13 ; 2431/23
1324/2 ; 1324/2	2314/2 ; 3124/1	3214/12 ; 3214/12	4213/12 ; 3241/13
1342/3 ; 1423/2	2341/3 ; 4123/1	3241/13 ; 4213/12	4231/13 ; 4231/13
1423/2 ; 1342/3	2413/2 ; 3142/13	3412/2 ; 3412/2	4312/12 ; 3421/23
1432/23 ; 1432/23	2431/23 ; 4123/13	3421/23 ; 4312/12	4321/123 ; 4321/123

There are 7 tables with row and column sums 112:

100	010	100	010	001	001	001
010	100	001	001	100	010	001
002	002	011	101	011	101	110

Now $D(1,1,2) = \{1,2\}$; from the display, pairs π, π^{-1} with $D(\pi) \subseteq \{1,2\}$, $D(\pi^{-1}) \subseteq \{1,2\}$ are

1234,1234   1324,1324   2134,2134   2314,3124   3124,2314   3214,3214   3412,3412.

Proof. of the theorem. Consider two compositions, **r** and **c** of N. A permutation π can be represented by a permutation matrix

$$\rho(\pi)_{ij} = \begin{cases} 1 & \text{if } \pi(i) = j \\ 0 & \text{otherwise.} \end{cases}$$

Thus, the 1 in the ith row is in position $\pi(i)$ and $\pi(i+1) < \pi(i)$ says the 1 in the $i+1$st row occurs to the left of the one in the ith row. Divide ρ into blocks specified by **r** and **c**. Then, π has $D(\pi) \subseteq \{r_1, r_1 + r_2, \cdots, r_1 + \cdots + r_{m-1}\}$ if and only if the pattern of ones in each horizontal strip decreases from upper left to lower right. Since $\rho(\pi^{-1}) = \rho(\pi)^t$, $D(\pi^{-1}) \subset \{c_1, c_1 + c_2, \cdots, c_1 + \cdots + c_{n-1}\}$ if and only if the pattern of ones in each vertical strip decreases from upper left to lower right.

With this representation, there is a one-to-one correspondence between tables $T \in \Sigma_{\mathbf{rc}}$ and permutations satisfying the constraints: form a permutation matrix with T_{ij} ones in the (i,j) block which also satisfies the monotonicity constraints. There is a unique way to do this: the T_{11} ones in the $(1,1)$ block must be contiguous along the diagonal, starting at $(1,1)$. The T_{12} ones in the $(1,2)$ block must be contiguous on a diagonal starting at $(T_{11} + 1, c_1 + 1)$; that is as far to the left and high up as possible consistent with the monotonicity constraints. The first horizontal block of $\rho(\pi)$ is similarly specified. Now the entries in the $(2,1)$ block and then the second horizontal strip are forced and so on. Continuing, we see that $\rho(\pi)$ is uniquely determined by T. $\square$

Foulkes (1976) and Garsia and Remmel (1985, pp. 233–234) give related results in terms of Schur functions. Briefly, let $c^{\nu}_{\lambda,\mu}$ be the Littlewood-Richardson numbers. These may be defined by $s_\lambda s_\mu = \Sigma_\nu c^{\nu}_{\lambda,\mu} s_\nu$ where s_λ, s_μ, s_ν are Schur functions. To connect these to descents, let D be a subset of $\{1, 2, \cdots, n-1\}$. Construct a skew diagram $\nu(D)/\lambda(D)$ consisting of a connected string of n boxes starting in the first column and ending in the first row, by moving right at step i if $i \in D$ and up if $i \in D^c$. For example, $D = \{1, 2, 4, 6, 7\}$ corresponds to the skew diagram :

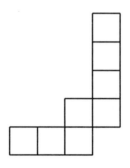

FIG. 3.1.

Foulkes showed that the number of permutations with descent set D whose inverse has descent set E is

$$\sum_\mu c^{\nu(D)}_{\lambda(D)\mu} c^{\nu(E)}_{\lambda(E)\mu}.$$

Because of the theorem above, summing this expression in D and E gives an expression for $|\Sigma_{\mathbf{rc}}|$.

3.2. Double cosets. Given a partition $\mathbf{r}$ of N, let $S_{\mathbf{r}}$ be the subgroup of the symmetric group S_N that permutes the first r_1 elements among themselves, the next r_2 elements among themselves, and so on. This $S_{\mathbf{r}}$ is called a Young subgroup and is a basic tool in developing the representation theory of S_N. It is isomorphic to the direct product of the S_{r_i}. Two Young subgroups $S_{\mathbf{r}}$ and $S_{\mathbf{c}}$ can be used to define double cosets. These are equivalence classes for the following relation:

$$\pi \sim \sigma \quad \text{if} \quad \rho\pi\kappa = \sigma \quad \text{for some } \rho \in S_{\mathbf{r}}, \ \kappa \in S_{\mathbf{c}}.$$

The following lemma is a classical combinatorial fact:

LEMMA 3.3. *In the symmetric group S_N, the number of double cosets for $S_{\mathbf{r}}, S_{\mathbf{c}}$ equals $|\Sigma_{\mathbf{rc}}|$.*

Proof. The correspondence between tables and cosets has the following combinatorial interpretation: consider N balls labeled 1 up to N. Color the first r_1 balls with color 1, the next r_2 balls with color 2, and so on. Let $\pi \in S_N$ permute the labels. Construct a table $T(\pi)$ as follows: look at the first c_1 places in π and for each color i count how many balls of color i occur in these places. Call these numbers $T(\pi)_{i,1}$. Then consider the next c_2 places in π and count how many balls of each color i occur. Call these numbers $T(\pi)_{i,2}$. Continuing gives a table $T(\pi) \in \Sigma_{\mathbf{rc}}$. It is not hard to check that every table in $\Sigma_{\mathbf{rc}}$ is $T(\pi)$ for some π and that $T(\pi) = T(\sigma)$ if and only if π and σ are in the same double coset. Thus the number of double cosets equals $|\Sigma_{\mathbf{rc}}|$. □

A group-theoretic proof of the lemma is given by James and Kerber (1981, Cor. 1.3.11). The above proof clearly shows that $\Sigma_{\mathbf{rc}}$ is nonempty.

3.3. Induced representations and tensor products. Let G be a finite group; $H \subseteq G$ a subgroup, $X = G/H$ the associated coset space, and $L(X)$ the vector space of all real valued functions on X. The group G acts on $L(X)$ by left translation: $sf(x) = f(s^{-1}x)$. The resulting representation is denoted $Ind_H^G(triv)$: the representation of G induced from the trivial representation of H. For $G = S_N$, with Young subgroup $S_{\mathbf{r}}$, these representations arise in the statistical analysis of "partially ranked data of shape $\mathbf{r}$". See Diaconis (1988, 1989). They are also the building blocks of most constructions of the irreducible representations of S_N.

For $\mathbf{r}$ and $\mathbf{c}$ partitions of N a classical theorem of Mackey (see, e.g., James and Kerber (1981, p. 17)) studies the intertwining number $I(\mathbf{r}, \mathbf{c})$: the dimension of the space of linear maps from $Ind_{S_{\mathbf{r}}}^{S_N}(triv)$ to $Ind_{S_{\mathbf{c}}}^{S_N}(triv)$ that commute with the action of S_N.

THEOREM 3.4 (MACKEY). *For $\mathbf{r}$ and $\mathbf{c}$ partitions of N,*

$$I(\mathbf{r}, \mathbf{c}) = |\Sigma_{\mathbf{rc}}|.$$

There is a related appearance. Let (ρ_1, V_1) and (ρ_2, V_2) be linear representations of a finite group G. The tensor product is the set of matrices $\rho_1(s) \otimes \rho_2(s)$ for $s \in G$. This gives a representation of G on $V_1 \otimes V_2$. One of the basic problems of representation theory is the study of how tensor products decompose.

If $M^{\mathbf{r}} = Ind_{S_{\mathbf{r}}}^{S_N}(triv)$, one can state

THEOREM 3.5 (MACKEY). *For $\mathbf{r}$ and $\mathbf{c}$ partitions of N*

$$M^{\mathbf{r}} \otimes M^{\mathbf{c}} = \bigoplus_T M^S$$

where the sum runs over all tables T in $\Sigma_{\mathbf{rc}}$ and the partition $S = S(T)$ is derived from T by taking all the entries of T in order.

For example, take $\mathbf{r} = \mathbf{c} = (N - 1, 1)$. There are two tables in $\Sigma_{\mathbf{rc}}$:

$$
\begin{array}{cc}
N-1 & 0 \\
0 & 1
\end{array}
\qquad
\begin{array}{cc}
N-2 & 1 \\
1 & 0
\end{array}.
$$

Thus

$$M^{N-1,1} \otimes M^{N-1,1} = M^{N-1,1} \oplus M^{N-2,1,1}.$$

This theorem is proved by James and Kerber (1981, pp. 95–98). It gives a convenient way of decomposing tensor products of irreducible representations as well. There is a general interrelation between double cosets, induced representations and tensor products which includes these results as a special case. Curtis and Reiner (1962) develop this in some detail. The special case has fascinating ramifications not developed here. This concerns Solomon's descent algebras which connect to Lie theory, card shuffling and much else. See Solomon (1976), Garsia (1990), Garsia and Reutenauer (1989) and Diaconis, McGrath, Pitman (1993).

4. Symmetric functions. A polynomial in n variables is called symmetric if it is invariant under every permutation of its variables. Let Λ_n be the ring of symmetric polynomials in variables $x_1, x_2, \cdots, x_n$, with integer coefficients. This ring decomposes into a direct sum of subrings

$$\Lambda_n = \bigoplus_{k \geq 0} \Lambda_n^k$$

where Λ_n^k consists of the homogeneous symmetric polynomials of degree k, together with the zero polynomial. The best reference for these matters is Macdonald (1979). We use his notation in this section.

There are a number of well known bases for Λ_n^k. All are indexed by partitions λ of k with n or fewer parts: $\lambda = (\lambda_1, \cdots, \lambda_n)$, $\lambda_1 + \cdots + \lambda_n = k$, $\lambda_i \geq 0$. Let $x^\lambda = x_1^{\lambda_1} \cdots x_n^{\lambda_n}$.

- Monomial symmetric functions m_λ are defined by

$$m_\lambda(x_1, \cdots, x_n) = \Sigma x^\alpha$$

summed over all distinct permutations of $(\lambda_1, \cdots, \lambda_n)$. Thus, if $n = 3, k = 4, \lambda = (2, 1, 1)$, $m_{211}(x_1, x_2, x_3) = x_1^2 x_2 x_3 + x_1 x_2^2 x_3 + x_1 x_2 x_3^2$. These m_λ are symmetric and form a basis for Λ_n^k.

- Elementary symmetric functions e_λ are defined by

$$e_j = \sum_{1 \leq i_1 < \cdots < i_j \leq n} x_{i_1} x_{i_2} \cdots x_{i_j} \quad \text{and} \quad e_\lambda = e_{\lambda_1} e_{\lambda_2} \cdots e_{\lambda_n}.$$

Thus $e_{211}(x_1, x_2, x_3) = (x_1 x_2 + x_2 x_3 + x_1 x_3)(x_1 + x_2 + x_3)^2$. These also form a basis for Λ_n^k.

- The complete symmetric functions h_λ are defined by

$$h_j = \sum_{1 \leq i_1 \leq \cdots \leq i_j \leq n} x_{i_1} x_{i_2} \cdots x_{i_j} \quad \text{and} \quad h_\lambda = h_{\lambda_1} h_{\lambda_2} \cdots h_{\lambda_n}.$$

Thus $h_{211}(x_1, x_2, x_3) = (x_1^2 + x_2^2 + x_3^2 + x_1x_2 + x_1x_3 + x_2x_3)(x_1 + x_2 + x_3)^2$. These again form a basis for Λ_n^k.

- The power sum symmetric functions are defined by

$$p_j = \sum_i x_i^j, \qquad p_\lambda = p_{\lambda_1} \cdots p_{\lambda_n}.$$

Thus $p_{211}(x_1, x_2, x_3) = (x_1^2 + x_2^2 + x_3^2)(x_1 + x_2 + x_3)^2$. These form a basis for Λ_n^k over $\mathbb{Q}$ but not over $\mathbb{Z}$: $h_2 = \frac{1}{2}(p_1^2 + p_2)$.

A scalar product can be defined on Λ_n^k by requiring that h and m are dual:

$$\langle h_\lambda | m_\mu \rangle = \delta_{\lambda\mu}$$

for δ the Kronecker delta. With this choice, the p_λ bases are orthogonal:

$$\langle p_\lambda | p_\mu \rangle = \delta_{\lambda\mu} z_\lambda,$$

where z_λ is defined in terms of the partition $\lambda = 1^{a_1} 2^{a_2} \cdots k^{a_k}$, by

$$z_\lambda = \prod_{i=1}^{k} i^{a_i} a_i! \, .$$

The following well known result connects all of this to arrays.

THEOREM 4.1. *If* $\mathbf{r}$ *and* $\mathbf{c}$ *are partitions of* N,

$$\langle h_{\mathbf{r}} | h_{\mathbf{c}} \rangle = |\Sigma_{\mathbf{rc}}|.$$

Proof. The transition matrix between the dual bases h and m is given by Macdonald as

$$h_{\mathbf{r}} = \sum_{\mathbf{c}} |\Sigma_{\mathbf{rc}}| m_{\mathbf{c}}.$$

Taking the inner product on both sides with $h_{\mathbf{c}}$ completes the argument. $\square$

We can draw two consequences from this result. The first is an algorithm for computing $|\Sigma_{\mathbf{rc}}|$. The second is a formula for this number.

The theorem offers a variety of schemes for computing $|\Sigma_{\mathbf{rc}}|$ by using currently available algorithms for computing with symmetric functions. Perhaps the best available tool is John Stembridge's package which runs in connection with Maple. An algorithm can be based on the identities above together with the following result (Macdonald (1979, p. 17)) which expresses the relation between the h and m bases:

$$h_j = \sum_{\lambda \vdash j} p_\lambda / z_\lambda.$$

Algorithm 4.2. (*to compute* $|\Sigma_{\mathbf{rc}}|$). Let $\mathbf{r}, \mathbf{c}$ be given. Express $h_{\mathbf{r}} = \prod_{i=1}^{m} h_{r_i}$ and $h_{\mathbf{c}}$ in terms of power sums. Then compute the inner product $\langle h_{\mathbf{r}} | h_{\mathbf{c}} \rangle$. Note that all terms involved in this algorithm involve positive quantities. Thus, truncation at any stage gives a lower bound.

As an example: there are two 2×2 tables with row and column sums $(2,1)$: $\begin{smallmatrix} 2 & 0 \\ 0 & 1 \end{smallmatrix}$ and $\begin{smallmatrix} 1 & 1 \\ 1 & 0 \end{smallmatrix}$. To compute $\langle h_{21} | h_{21} \rangle$, express $h_2 = \frac{1}{2}(p_2 + p_1^2)$, $h_1 = p_1$ and then $\langle h_{21} | h_{21} \rangle = \frac{1}{4}\langle (p_2 + p_1^2)p_1 | (p_2 + p_1^2)p_1 \rangle = \frac{1}{4}\{\langle (p_2 p_1) | p_2 p_1 \rangle + \langle p_1^3 | p_3 p_1 \rangle \} = \frac{1}{4}\{2 + 6\} = 2$.

Alas, in practice, computation does not seem so feasible for N above 100 or so. As will emerge, other algorithms work well with much larger N. As such N arise in practice, improvements in symmetric function technology are needed before algorithm (4.2) becomes feasible.

Just carrying through the algorithm gives the following formula:

COROLLARY 4.3. *Let* $\mathbf{r}$ *and* $\mathbf{c}$ *be partitions of* N. *Then*

$$|\Sigma_{\mathbf{rc}}| = \sum_{\substack{\mu^i \vdash r_i \\ \lambda^j \vdash c_j}} \delta_{\lambda\mu} z_\lambda \prod \frac{1}{z_{\mu_i} z_{\lambda_j}}.$$

The sum is over all partitions μ^i *of* r_i, λ^j *of* c_j *for* $1 \leq i \leq m$, $1 \leq j \leq n$ *and* $\lambda = (\lambda^1, \cdots, \lambda^m)$, $\mu = (\mu^1, \cdots, \mu^n)$ *are considered partitions of* N *by concatenating parts.*

Remark 4.4. In Section 7.2 we give the generating function for the number of tables as $\prod_{i,j}(1 - x_i y_j)^{-1} = e^{\Sigma p_i(x)p_i(y)/i}$. On the right, $p_i(x)$ is the power sum symmetric function. The corollary can also be read of this expansion by expanding the exponential in the usual way. Richard Stanley used this technique to get numerical answers to problem 27974 in the American Mathematical Monthly (1980).

5. Young tableaux and Kostka numbers. For λ and μ partitions of n, a *semi-standard Young tableau of shape* λ *and content* μ is a diagram of shape λ containing μ_1 ones, μ_2 twos, etc., arranged to be weakly increasing in rows and strictly increasing down columns. For example,

$$\begin{matrix} 1 & 1 & 2 \\ 2 & 3 & \end{matrix}$$

is semi-standard with shape 3,2 and content 2,2,1. Such tableaux are a basic ingredient in the description of the irreducible representations of the classical groups.

Define the Kostka number $K_{\lambda\mu}$ as the number of semi-standard tableaux of shape λ and content μ. The following classical result relates these to arrays.

THEOREM 5.1. *Let* $\mathbf{r}$ *and* $\mathbf{c}$ *be partitions of* N. *Then*

$$\left| \Sigma_{\mathbf{rc}} \right| = \sum_{\mu} K_{\mathbf{r}\mu} K_{\mathbf{c}\mu}.$$

Proof. The Schur functions s_λ are yet another basis for the ring Λ_n^k of Section 4. These can be expressed in terms of the h_λ as

$$h_\lambda = \sum_{\mu} K_{\lambda\mu} s_\mu.$$

Macdonald (1979, p. 57) proves this as well as showing that $\langle s_\lambda | s_\mu \rangle = \delta_{\lambda\mu}$. Now $\left| \Sigma_{\mathbf{rc}} \right| = \langle h_\mathbf{r} | h_\mathbf{c} \rangle$, so the result follows. $\square$

Remark 5.2. 1. A direct combinatorial proof of the theorem is given by Knuth (1970). Extending ideas of Robinson, Schensted, and Schützenberger, he gives a bijection between tables with row sum $\mathbf{r}$ and column sum $\mathbf{c}$ and pairs of semi-standard tableaux with shape $\mathbf{r}$ and content $\mathbf{c}$.

2. There has been some recent work on formulae for the Kostka numbers $K_{\lambda\mu}$. At the moment, these do not seem so useful, but here is a brief description. Kirillov and Reshitikhin (1986) have shown

$$K_{\lambda\mu} = \sum_{\alpha} \prod_{k,n \geq 1} [p_n^k(\alpha), \alpha_n^k - \alpha_{n+1}]$$

where the sum is over all sequences of partitions $\alpha = (\alpha^0, \alpha^1, \alpha^2 \cdots)$ such that $\alpha^0 = \mu$, $|\alpha^k| = \lambda_{k+1} + \lambda_{k+2} + \cdots$, $k \geq 1$, further $p_n^k(\alpha) = \sum_{i=1}^{n}(\alpha_i^{k-1} - 2\alpha_i^k + \alpha_i^{k+1})$ and $[a, b] = \begin{pmatrix} a + b \\ b \end{pmatrix}$.

6. Tables and statistics. This section motivates statistical uses of arrays and explains how enumeration of $\Sigma_{\mathbf{rc}}$ arises in application. Data is often categorized into 2-way contingency tables. The following example is typical: 592 subjects were classified by hair and eye color (Snee (1974)). Such data are often analyzed under the assumption that the cell counts T_{ij} follow a multinomial distribution with probability p_{ij}. The independence hypothesis can be specified as

$$p_{ij} = p_{i \cdot} p_{\cdot j} \quad \text{for all} \quad i \quad \text{and} \quad j, \quad \text{where} \quad p_{i \cdot} = \sum_{j} p_{ij}, \quad p_{\cdot j} = \sum_{i} p_{ij}.$$

A standard test of independence uses the chi-squared statistic

$$\chi^2 = \sum_{i,j} \frac{(T_{ij} - \frac{r_i c_j}{n})^2}{\frac{r_i c_j}{n}}, \quad \text{with} \quad r_i = \sum_{j} T_{ij}, \quad c_j = \sum_{i} T_{ij}.$$

In Table 6.1, $\chi^2 = 138.29$.

TABLE 6.1

	Black	Brunette	Red	Blonde	
Brown	68	119	26	7	220
Blue	20	84	17	94	215
Hazel	15	54	14	10	93
Green	5	29	14	16	64
	108	286	71	127	592

The chi-squared value is usually compared with an approximation from the chi-squared distribution. In this example, the approximation has mean 9 and standard deviation $\sqrt{18}$, so 138.29 is a huge value and independence is rejected.

Huge values of chi-squared are sufficiently common that statisticians have developed other ways of calibrating the chi-square statistic. For example, Diaconis and Efron (1985) assumed that the underlying probabilities p_{ij} were unknown and put a uniform prior on them. This is just Lebesgue measure on the simplex $\{p_{ij} : p_{ij} \geq 0, \Sigma p_{ij} = 1\}$. Under this assumption, it turns out that the table T_{ij} is also uniform: all tables have an equal chance of occurring. This suggests calibrating the distribution of χ^2 under the uniform distribution as an antagonistic alternative to the model of independence.

In statistics, it is customary to fix (or condition on) the row and column sums of the observed table and ask for calibration of test statistics in the set of possible tables. This leads to the following combinatorial problem: for $\mathbf{r}$ and $\mathbf{c}$ partitions of N, find the proportion of tables in $\Sigma_{\mathbf{rc}}$ with chi-squared values smaller than t, as t varies.

Diaconis and Efron (1985) develop a variety of techniques to approximate this distribution and describe related work by Good and other statisticians. A Monte Carlo algorithm discussed in Section 10 below suggests that in fact, about 15.4% of all tables have $\chi^2 \leq 138.29$. Thus here the data is compatible with the antagonistic alternative.

Enumeration of $\Sigma_{\mathbf{rc}}$ enters the picture at several points. At present, the only way we have of exactly solving the calibration problem under the uniform distribution on $\Sigma_{\mathbf{rc}}$ is to systematically run through all tables and actually calculate the statistic. It is of obvious interest to have an estimate of the number of tables to have some idea of the running time.

The size of $|\Sigma_{\mathbf{rc}}|$ is similarly used to estimate the running time of algorithms for exact enumeration of the chi-squared statistic under other distributions on $\Sigma_{\mathbf{rc}}$. The most important of these is the Fisher-Yates (or multiple hypergeometric) distribution. A variety of algorithms for doing these computations are reviewed in Sections 8 and 9.

Finally, the size of $|\Sigma_{\mathbf{rc}}|$ was required as input to an approximation procedure proposed by Diaconis and Efron (1985).

In Section 8.1 we give algorithms for calculating $|\Sigma_{\mathbf{rc}}|$ which work for tables like Table 6.1: there are approximately 10^{15} tables with the same row and column sums as Table 6.1; in fact, there are $1,225,914,276,768,514$ such tables.

We conclude this section by mentioning some literature related to privacy issues. Suppose a contingency table is summarized by giving its row and column sums and perhaps a few other functions. How much about the individual cell entries can be deduced? Gusfield (1988) gives best possible bounds on the entries and a review of this literature.

7. Asymptotic approximations.

A variety of asymptotic approximations to $|\Sigma_{\mathbf{r,c}}|$ have been suggested (and proved!). O'Neil (1969), Békéssy, Békéssy and Komlós (1972) followed by Good and Crook (1977) give

$$(7.1) \qquad |\Sigma_{\mathbf{rc}}| \sim \frac{N!}{\Pi r_i! \Pi c_j!} \exp\left\{ \frac{2}{N^2} \sum_{i,j} \binom{r_i}{2}\binom{c_j}{2} \right\}.$$

O'Neil proved this as m, n tend to infinity with the row sums of less than $\{\log \max(m, n)\}^{1/4-\epsilon}$; that is for large, sparse tables. Bender (1974) developed variations for prescribed zeros and bounded entries. Békéssy, Békéssy and Komlós proved it was valid for m fixed and n large. Alas, this approximation is useless for tables like Table 6.1, 8.1 above where both m and n are small and N is large. They are off by many orders of magnitude for these examples.

Diaconis and Efron (1986) gave approximations which seem to work for m, n small, N large. To state the result, let

$$w = \frac{1}{1 + mn/2N}, \quad k = \frac{n+1}{n\Sigma \bar{r}_i^2} - \frac{1}{n}, \quad \bar{r}_i = \frac{1-w}{m} + \frac{wr_i}{N}, \quad \bar{c}_j = \frac{1-w}{n} + \frac{wc_j}{N}.$$

Then Diaconis and Efron suggest (without proof)

$$|\Sigma_{\mathbf{rc}}| \sim \left(\frac{2N+mn}{2} \right)^{(m-1)(n-1)} \left(\prod_{i=1}^{m} \bar{r}_i \right)^{n-1} \left(\prod_{j=1}^{m} \bar{c}_i \right)^{k-1} \frac{\Gamma(nk)}{\Gamma(n)^m(k)^n}.$$

(7.2)

For example, this gives 1.235×10^{15} as an approximation of the number of tables with the same margins as Table 6.1. The right answer is 1.226×10^{15}. For Table 8.1 it gives 2.33×10^8, the right answer being 2.39×10^8.

Good and Crook (1977) suggest several further approximations the simplest of these, (6.5) of their paper, gives 1.1432×10^{15}. For Table 6.1 and the remarkable 2.3939×10^8 for table 8.1.

Finally, Gail and Mantel (1977) have suggested a normal approximation which we have not found terribly reliable.

8. Algorithms for exact enumeration. This section describes many algorithms which have been developed for counting or actually running through all $m \times n$ tables with row sum $\mathbf{r}$ and column sums $\mathbf{c}$. For the $2 \times n$ and $3 \times n$ case, formulas of Mann are described in Section 8.4 below. Algorithms using symmetric functions were described in Sections 4 and 5. Of course, the space of tables $\Sigma_{\mathbf{rc}}$ is large and the complexity considerations of Section 9 may rule out success. Nonetheless, we have been pleasantly surprised at how often exact answers are available for problems of practical interest.

8.1. Exact enumeration. A number of authors, March (1972), Boulton and Wallace (1973), Hancock (1974), Balmer (1988) have suggested algorithms for exhaustively stepping through the set $\Sigma_{\mathbf{rc}}$ one table at a time. These begin at some canonically constructed initial table and proceed by making small changes to the cell entries so that the tables increase monotonically in some linear order. This affords a means of calculating $|\Sigma_{\mathbf{rc}}|$ as well as various other functions on the set such as the proportion of tables with a chi-square (or other statistic) smaller than a given value.

As an example, Table 8.1 below was chosen as a table with irregular margins. There are 239,382,173 tables in $\Sigma_{\mathbf{rc}}$. The exhaustive enumeration

TABLE 8.1

5	2	3	10
50	7	5	62
3	6	4	13
5	3	3	11
2	7	30	39
65	25	45	135

using Pagano and Taylor-Halvorsen's algorithm on a DEC station 3100 (a vintage 1990 desktop workstation) took 54665 CPU seconds and 15 hours, 21 minutes of real time to calculate the exact chi-square distribution. John Mount reported that counting alone took 17 seconds using the divide and conquer ideas explained below.

Table 8.1 has $\chi^2 = 72.18$. A histogram of the χ^2 values for all tables in $\Sigma_{\mathbf{rc}}$ is shown in Figure 8.1. The exact proportion of tables with chi-square value smaller than 72.18 is .76086 to 5 significant figures. This will be used to illustrate some other approaches (Figure 8.1 is explained in Section 10.2.)

If one is interested in simply determining the proportion of tables with chi-square values smaller than a given value, one may follow R.A. Fisher (1935) and "work in from the end". This avoids generating every table.

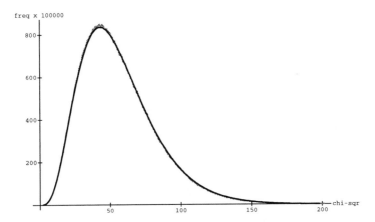

FIG. 8.1. *The dark curve shows the exact distribution of the* χ^2 *statistic. The light curve shows a Monte Carlo approximation using the random walk of section 10.2*

Pagano and Taylor-Halvorsen (1981) have developed such a short cut for the hypergeometric distribution on $\Sigma_{\mathbf{rc}}$.

One straightforward approach to table enumeration uses a recurrence. Gail and Mantell (1977) carry this out. For example, in the $m \times 3$ case, let $T(r_1, \cdots, r_m; c_1, c_2) = |\Sigma_{\mathbf{rc}}|$ then

$$T(\mathbf{r}, \mathbf{c}) = \sum_{k_1, k_2} T(r_1, r_2, \cdots, r_{m-1}; c_1 - k_1, c_2 - k_2).$$

The sum on the right runs over values k_1, k_2 with $0 \leq k_i \leq \min(r_m, c_i)$ and $k_1 + k_2 \leq r_m$. Of course, the generalization of this recurrence to arbitrary dimensions takes exponential time to compute.

We mention briefly two other algorithms which have seen extensive empirical application. Stein and Stein (1970) have proposed a branching algorithm. This is extended by Good and Crook (1977) who give a clear description. Finally, the network algorithm of Mehta and Patel (1983) is a mainstay of the commercial program Statexact. It is geared for exact evaluation of hypergeometric distributions for statistics on $\Sigma_{\mathbf{rc}}$.

Finally, we mention that David des Jardin (personal communication) and John Mount (personal communication) have employed a divide and conquer algorithm. Mount's algorithm works by running through all possible $2 \times n$ tables T with column sums $\mathbf{c}$ and row sums $\mathbf{r}'$, where $r'_1 = r_1 + \cdots + r_{\lfloor \frac{m}{2} \rfloor}$, $r'_2 = r_{\lfloor \frac{m}{2} \rfloor + 1} + \cdots + r_m$. Let $\mathbf{r}^\ell$ be the $\lfloor \frac{m}{2} \rfloor$ vector consisting of the first $\lfloor \frac{m}{2} \rfloor$ entries of $\mathbf{r}$ and $\mathbf{r}^R$ be the $m - \lfloor \frac{m}{2} \rfloor$ vector consisting of the last entries of $\mathbf{r}$. Let T_i denote the ith row of T. Then

$$m(\mathbf{r}, \mathbf{c}) = \Sigma_T m(\mathbf{r}^\ell, T_1) \times m(\mathbf{r}^\ell, T_2),$$

where T runs through all legal $2 \times n$ tables having row sums $\mathbf{r}'$ and column sums **c**. The smaller counting problems are solved by the same technique. The recursion ends when a $1 \times n$ or $2 \times n$ problem is reached. This algorithm has solved the hardest problems to date, the 4×4 example in Table 6.1, which has $1,225,914,276,768,514$ tables with the same row and column sums.

8.2. Generating functions and Fourier transforms.

Let $x_1, x_2, \cdots, x_m$ and $y_1, y_2, \cdots, y_n$ be variables. Form the generating function

$$\prod_{i,j}(1 - x_i y_j)^{-1} = (1 + x_1 y_1 + (x_1 y_1)^2 + \cdots)(1 + x_1 y_2 + (x_1 y_2)^2 + \cdots)$$
$$\cdots (1 + x_m y_m + (x_m y_m)^2 + \cdots).$$

(8.1)

By inspection, the coefficient of $x_1^{r_1} x_2^{r_2} \cdots x_m^{r_m} y_1^{c_1} y_2^{c_2} \cdots y_n^{c_n}$ is $|\Sigma_{\mathbf{rc}}|$. For example, the coefficient of $x_1^2 x_2 y_1 y_2 y_3$ is 3. The 3 tables with $\mathbf{r} = (2, 1), \mathbf{c} = (1, 1, 1)$ are

$$\begin{matrix} 1 & 1 & 0 \\ 0 & 0 & 1 \end{matrix} \qquad \begin{matrix} 1 & 0 & 1 \\ 0 & 1 & 0 \end{matrix} \qquad \begin{matrix} 0 & 1 & 1 \\ 1 & 0 & 0 \end{matrix}.$$

The coefficients in the generating function can be expressed as contour integrals and one can attempt asymptotic approximations. Good (1976) pursues this line; see Section 9.

Along more algorithmic lines, we can truncate the expansions above and compute the initial portions of the product by multiplying polynomials. This can in turn be done using fast Fourier transform (F.F.T.)

We outline the method and study its running time. Let $\mathbf{i} = (i_1, i_2, \cdots, i_k)$ and let $z^{\mathbf{i}} = z_1^{i_1} \cdots z_k^{i_k}$. A polynomial is $f(z_1, \cdots, z_k) = f(z) = \Sigma_{\mathbf{i}} a_{\mathbf{i}} z^{\mathbf{i}}$. Two k variable polynomials of degree at most d in each variable can be multiplied by using the F.F.T. on the group $\mathbf{Z}_{2d}^k$ in $O(k(2d)^k \log d)$ operations. To see this, suppose $f(z) = \Sigma_{\mathbf{i}} a_{\mathbf{i}} z^{\mathbf{i}}$ and $g(z) = \Sigma_{\mathbf{i}} b_{\mathbf{i}} z^{\mathbf{i}}$. Their product is $h(z) = \Sigma_{\mathbf{k}} c_{\mathbf{k}} z^{\mathbf{k}}$ where $c_{\mathbf{k}} = \Sigma_{\mathbf{i}+\mathbf{j}=\mathbf{k}} a_{\mathbf{i}} b_{\mathbf{j}}$. Such convolutions can be computed by the F.F.T. on $\mathbf{Z}_{2d}^k$ in $O(k(2d)^k \log d)$ operations. One useful reference is Cormen, Leiserson, Rivest (1990, Chapter 32).

Moving back to tables, let $d_* = \max\{r_i, c_j, 1 \le i \le m, 1 \le j \le n\}$. The coefficient of the term $x^{\mathbf{r}} y^{\mathbf{c}}$ in the generating function (1.1) cannot depend on any term in the product with degree exceeding d_*. Successively multiplying each of the mn polynomials and discarding terms of excess degree gives the result we want. We summarize:

LEMMA 8.1. *Given vectors* $\mathbf{r}$ *and* $\mathbf{c}$, *of length* m *and* n, *let* $d_* = \max\{r_i, c_j, 1 \le i \le m, 1 \le j \le n\}$. *There is an algorithm for computing* $|\Sigma_{\mathbf{rc}}|$ *using* $O(mn(m+n)(2d_*)^{m+n} \log d_*)$ *operations on* $0((2d_*)^k)$ *numbers.*

Remark 8.2. 1. For fixed m and n this gives an algorithm for computing $|\Sigma_{\mathbf{rc}}|$ whose running time is a polynomial in N. Similar ideas are suggested by Good and Crook (1977).

2.   The polynomial technique can be adapted to count elements T in $\Sigma_{\mathbf{rc}}$ which satisfy additional linear constraints of the form

$$\sum_{i,j} a_{ij} T_{ij} = a.$$

This can be done by using the generating function

$$\prod_{i,j} (1 - s^{a_{ij}} x_i y_j)^{-1}.$$

The coefficient of $s^a x^r y^c$ gives the number of tables in $\Sigma_{\mathbf{rc}}$ satisfying the constraint. Any number of constraints can be handled this way.

For example, the generating function for $n \times n$ "magic squares" with diagonal sums equal to row and column sums can be expressed as

$$\prod_{i,j} (1 - s^{a_{ij}} t^{b_{ij}} x_i y_j)^{-1},$$

where $a_{ij} = 1$ if $i = j$ and zero otherwise, and $b_{ij} = 1$ if $i + j = n$ and zero otherwise.

Such additional constraints arise naturally in contingency table analysis. For example, Agresti, Mehta and Patel (1990) needed the number of tables with prescribed row and column sums and an additional constraint as above with $a_{ij} = u_i v_j$ for specific u_i, v_j.

The computational approach centered around polynomials has been actively developed by workers in computational statistics. Baglivio (1994) gives a book length development centered around this theme. She gives an extensive review of the statistical literature.

8.3. Ehrhart polynomials and toric ideals. There has been an active recent development in combinatorial mathematics, commutative algebra, and algebraic geometry which leads to useful algorithms for table enumeration. Briefly, the number of tables with given row and column sums can be shown to be a piecewise polynomial in $\mathbf{r}$ and $\mathbf{c}$ which shifts its coefficients at boundaries specified by well-specified hyperplanes. Further, the polynomials can be identified by using geometric properties of an associated "toric ideal". The following example was kindly communicated by Bernd Sturmfels.

Let $T(r_2, r_3, r_4, c_1, c_2, c_3, c_4)$ be the number of 4×4 tables with row sums r_1, r_2, r_3, r_4 and column and sums c_1, c_2, c_3, c_4, where $r_1 = c_1 + c_2 + c_3 + c_4 - r_2 - r_3 - r_4$. Consider the following closed convex polyhedral cone $\mathcal{C}$ which is defined by the following 11 linear inequalities in the 7-dimensional space of marginal totals:

$c_1 \geq r_3, c_4 \geq r_3, c_3 \geq r_4, c_1 + c_4 \geq r_2, r_3 + r_4 \geq c_4, r_3 + r_4 \geq c_1, c_2 \geq r_2 + r_4$

$r_2 + r_4 \geq c_1 + c_4, c_2 + c_3 \geq r_2 + r_3, r_2 + r_3 \geq c_1 + c_3 + c_4, r_2 + r_3 + r_4 \geq c_2 - c_3.$

This cone has 18 extreme rays and it contains the marginal totals of the data in Table 6.1 in its interior: $(215, 93, 64, 108, 286, 71, 127) \in Int(\mathcal{C})$. Sturmfels has proved that the restriction of T to $\mathcal{C}$ is a polynomial with rational coefficients of degree 9. As an example of what can be done, suppose that c_4 is varied by δ. How does that effect the number of tables? The polynomial specializes to the following formula which is valid for $-20 \leq \delta \leq 2$:

$$T(215, 93, 64, 108, 286, 71, 127 + \delta) =$$

$$-\frac{1}{40320}\delta^9 - \frac{37}{5040}\delta^8 + \frac{1871}{2240}\delta^7 + \frac{5269}{36}\delta^6 - \frac{60566509}{1920}\delta^5 - \frac{587569069}{720}\delta^4$$

$$-\frac{1965953600567}{10080}\delta^3 - \frac{485892234635}{84}\delta^2 + \frac{105240591524160}{7}\delta + 1225914276768514.$$

Of course, when $\delta = 0$ we recover the number of tables from the final coefficient.

That T is a piecewise polynomial follows from the theory developed by E. Ehrhart. The clearest elementary treatment is in Stanley (1986, Sections 4.4, 4.6). This contains background and references. The quasi polynomial nature of the answer suggests an exciting possibility. One can determine the polynomial by computing with tiny marginal totals in the same part of the cone. This is how the magic square polynomials of Section 2 were determined.

The methods used to find the polynomial above lie somewhat deeper. Briefly, consider a set of integer vectors in $\mathbb{R}^d$. Form their convex hull $\mathcal{H}$ and consider the problem of finding N the number of lattice points in $\mathcal{H}$. Clearly the problem of table enumeration fits into this mold. Associated to $\mathcal{H}$ is a complex variety called a toric variety. There is a formula for N in terms of the geometry of this toric variety. It uses fairly abstract constructions such as the cohomology and Todd classes associated to the variety. Further, one can actually calculate the numerical ingredients of these geometric quantities. It would take us too far afield to develop these topics here. Danilov (1978) or Fulton (1993) give fine introductions which get to the relevant parts of the subject. Fulton gives references to recent work by Barvinock, Morelli, and Pomerschein on this subject.

8.4. Tables with fixed m and n. In applications, one is often in the situation in which many subjects are classified into a small number of categories. Table 6.1 of Section 6 is a typical example. We give a polynomial time algorithm (in N) for solving the problems in Section 8.2. Mann (1994) has given the following formulae for the case of $2 \times n$ or $3 \times n$ tables.

$$m_2(\mathbf{r}; \mathbf{c}) = \sum_{\sigma \subset [1,n]} \left[\begin{array}{c} r_1 - c_\sigma - |\sigma| \\ n - 1 \end{array} \right] (-1)^{|\sigma|}$$

$$m_3(\mathbf{r}; \mathbf{c}) = \left[\begin{array}{c} r_1 \\ n - 1 \end{array} \right] m_2(r_3, c_{[1,a]} - r_3; \mathbf{c})$$

$$\sum_{\substack{\sigma \subset [1,n] \\ |\sigma| \neq 0, n}} \sum_{\substack{\tau : \tau \cap \sigma = \emptyset \\ \gamma : \gamma \subset \sigma}} (-1)^{|\sigma| + |\tau| + |\gamma|} f$$

where

$$f = \sum_{k=0}^{|\sigma|-1} (-1)^k \left[\begin{array}{c} A + B - k \\ 2n - |\sigma| + k - 1 \end{array} \right] \left[\begin{array}{c} A + c_\gamma + |\gamma| - |\sigma| \\ |\sigma| - |\gamma \end{array} \right] \left[\begin{array}{c} n - 1 \\ k \end{array} \right]$$

$$+ (-1)^\sigma \sum_{j=1}^{n} \left[\begin{array}{c} n - j \\ |\sigma| - 1 \end{array} \right] \left[\begin{array}{c} A + c_\gamma + |\gamma| - |\sigma| \\ j - 1 \end{array} \right] \left[\begin{array}{c} B - c_\gamma - |\gamma \\ 2n - j - 1 \end{array} \right].$$

In these sums, $\left[\begin{array}{c} n \\ k \end{array} \right] = {}^{n+k}\!k$, $[1,n]$ is the interval from 1 to n; Greek letters denote subsets, and $c_\sigma = \sum_{j \in \sigma} c_j$. Finally, $A = r_1 - c_\sigma - |\sigma|$ and $B = r_3 - c_\tau - |\tau|$.

Similar formulae for the volume of the associated convex polyhedra appear in Diaconis and Efron (1985). For fixed n, these give a polynomial computation for $|\Sigma_{\mathbf{rc}}|$ in $\mathbf{r}$ and $\mathbf{c}$. The computation is also clearly exponential as n grows. The methods used to derive these formulae can be extended to ever more unsightly formulae for $m \times n$ tables. Dahmen and Micchelli (1988) relate such formulae to multivariate splines.

This puts the tension back in the problem; in a specific example, what is large? It seems feasible to us that there might also be tractable methods in sparse cases where m and n are large, but N is small or moderate.

9. Complexity results. This section records what is currently known about the complexity of computing $|\Sigma_{\mathbf{rc}}|$. There are 3 natural parameters that give a crude measure of problem size: m, n, and N. Briefly, the problem is hard (#-P complete) if m or n is large (theorem of Dyer, Kannan and Mount). If m and n are fixed and N is large, the problems are theoretically tractable. We begin with a brief self contained description of #-P completeness. The initiated may skip to Sections 9.2 and 9.3 which describe our results.

9.1. #-P completeness. The standard reference on this subject is Gary and Johnson (1979). One clear way to represent the relevant complexity classes begins with a finite alphabet Σ. Let Σ^* denote all finite sequences (words) of elements in Σ. We use $|x|$ to denote the length of x.

Let $R \subseteq \Sigma^* \times \Sigma^*$ be a relation on words. Write $R(x, y)$ to denote $(x, y) \in R$ and $R(x) = \{y : R(x, y)\}$: this is called the solution set of x. R is a *p-relation* if there are polynomials p and q such that:

- The predicate $R(x, y)$ can be checked in $p(|x|)$ operations.
- If $y \in R(x)$ then $|y| \leq q(|x|)$.

The well known class NP can be identified with the decision problems "Given x, is $R(x)$ non-empty?", where R is a P-relation. Similarly, the class $\#P$ can be identified with the counting problems "Given x what is the cardinality of $R(x)$?" where R is a P-relation.

Both in the case NP and $\#P$, there exist problems in the class that are complete under *polynomial time reductions*; these are problems ψ such that the existence of polynomial-time algorithms for ψ would imply the existence of polynomial-time algorithms for each problem in the class. Such problems are called NP complete or $\#P$ complete, respectively.

9.2. Some #-P completeness results for table enumeration. Dyer, Kannan and Mount (1994) have announced the result that for $m = 2$ and large n, N, the problem of determining $|\Sigma_{\mathbf{rc}}|$ is $\#P$-complete. We present here an earlier result of Gangolli (1991) which proves $\#$-P completeness for a practical class of problems.

Let Z be an $m \times n$ binary matrix. The set $\Sigma_{\mathbf{rc}}^Z$ of contingency tables with structural zeros at Z is the subset of $\Sigma_{\mathbf{rc}}$ in which every table has only zeros in positions where $Z(i, j) = 1_0$. This set arises naturally in the analysis of contingency tables where the row/column classification gives rise to forbidden combinations. For example, a table that classifies a population of subjects by sex (rows) and cause of death (columns) might have a forbidden entry representing males with uterine cancer. See Bishop, Fienberg and Holland (1975) for background.

THEOREM 9.1. *With* $\mathbf{r}, \mathbf{c}$ *and* Z *as parameters, the counting problem for* $\Sigma_{\mathbf{rc}}^Z$ *is #-P complete.*

Proof. First note that the counting function for the set $\Sigma_{\mathbf{rc}}^Z$ is in $\#$-P. To prove completeness, we give a reduction from the problem of computing the permanent. Here, if A is the $n \times n$ adjacency matrix of a bipartite graph on two sets of n vertices, *per* (A) is the number of perfect matchings; this is $\sum_\pi \prod_{i=1}^n A_{i\pi(i)}$ summed over the permutations of n. The problem of computing the permanent was shown to be $\#$-P complete by Valiant (1979).

Computing the permanent is a special case of computing $|\Sigma_{\mathbf{rc}}^Z|$. Given the $n \times n$ adjacency matrix A for a bipartite graph G let $\mathbf{r} = \mathbf{c} = (1, 1, \cdots, 1)$ (length n). Let $Z_{ij} = 1 - A_{ij}$. Now it is clear that a table T is in $\Sigma_{\mathbf{rc}}^Z$ if and only if T is the adjacency matrix of a perfect matching in G. Thus, $|\Sigma_{\mathbf{rc}}^Z|$ equals the number of perfect matchings. The reduction can be done in linear time and logarithmic space in the size of A. $\qquad\square$

Remark 9.2. The theorem holds even if the inputs are expressed in unary. Roughly speaking, this means that the difficulty of this problem is really due to the structure of the problem and not just the size of the marginal totals. See Garey and Johnson (1979) for further discussion.

Throughout, it is natural to inquire about the natural generalizations to three-dimensional arrays. The problems are surprisingly more difficult; even determining if the set of tables with given line sums is non-empty is NP complete. See Irving and Jerrum (1990).

10. Approximate counting using sampling. In this section we show that there are randomized algorithms that give accurate approximations to $|\Sigma_{\mathbf{rc}}|$ in a polynomial number of steps in m, n, N. The algorithms work by choosing tables at random and using these to approximate the number of tables. The conversion is explained first, followed by two methods of random choice—a combinatorial random walk and a convex set approach due to Dyer, Kannan and Mount.

10.1. Random walks on $\Sigma_{\mathbf{rc}}$. Fix row and column sums $\mathbf{r}, \mathbf{c}$. A variety of random walks on $\Sigma_{\mathbf{rc}}$ have been in active use by statisticians. The idea is simple: pick a pair of rows i, i' and a pair of columns j, j' uniformly at random. These rows and columns intersect in four entries. The walk proceeds by changing the current table into a new one by adding and subtracting one in these entries according to the following pattern:

$$\begin{matrix} + & - \\ - & + \end{matrix} \quad \text{or} \quad \begin{matrix} - & + \\ + & - \end{matrix}.$$

The final choice is made with probability $1/2$. If a step forces a negative table entry, the random walk stays at the original table. It is easy to see that this is a connected, symmetric, aperiodic Markov chain on $\Sigma_{\mathbf{rc}}$. It converges to the uniform distribution. A formal proof and discussion of rates of convergence appears in Section 10.2 below.

As an example, consider Table 8.1 of Section 8. This had a chi-square statistic of 72.18. A random walk was run on the tables with the same row and column sums in each run, the walk was run 49,936 as an initial randomization. Then, the walk was run 2,297,000 steps. For each step, the chi-square statistic was computed. A 1 is recorded if this is smaller than 72.18. The number of 1's/2,297,000 gives an estimate of the proportion of tables with chi-square smaller than 72.18. The whole procedure was repeated 5 times. The median of these five values is .7638.

In this example, an exhaustive enumeration gave the exact proportion as .76086 (to 5 significant figures) of the 239,382,173 tables. Thus, the Monte Carlo gives accurate answers for this example. The entire procedure took 23 minutes and 31 seconds. This is about 1/30 the time required for the exhaustive method. Figure 8.1 in Section 8 shows that the random walk gives a remarkably accurate approximation to the true distribution.

A history, many variations, and extensions of this algorithm are described by Diaconis and Sturmfels (1993). They also describe other problems (e.g., three-dimensional arrays) where similar algorithms are available.

10.2. From sampling to counting. This section gives a brief synopsis of work of Jerrum, Valiant, and Vazirani (1986). Sinclair (1993) gives a more complete short expository account.

If V is a finite set, H a subset and we can efficiently choose randomly from the uniform distribution over V, then we can estimate $|H|/|V|$ by seeing what proportion of samples fall into H. If H is small enough to be enumerated but large enough that $|H|/|V|$ is "not too small" (so the ratio can be reliably estimated) then the ingredients can be put together to give an estimate of $|V|$. In practice, a nested decreasing subsequence $V \supset H_1 \supset H_2 \cdots \supset H_r$ is used to meet the goals.

To make this rigorous, we will use the notation of Section 9. The definitions are a bit abstract; they are immediately followed by an example. Let Σ be a finite alphabet, Σ^* the finite sequences (or words) with each term in Σ. A relation $R \subset \Sigma^* \times \Sigma^*$ is *polynomially self-reducible* if

- There is a deterministic polynomial-time computable function $g : \Sigma^* \to \mathbb{N}$ such that if $R(x, y)$ then $|y| = g(x)$.
- There exist polynomial-time computable functions $\psi : \Sigma^* \times \Sigma^* \to \Sigma^*$ and $\sigma : \Sigma^* \to N$ such that for all $x, w \in \Sigma^*$

  $\sigma(x) \leq c \log |x|$ for some constant c,
  $g(x) > 0$ implies $\sigma(x) > 0$,
  $|\psi(x, w)| \leq |x|$,

  and such that for all $x \in \Sigma^*$, if $y = wz$ with $|y| = g(x)$ and $|w| = \sigma(x)$ then $R(x, wz)$ if and only if $R(\psi(x, w), z)$.

These conditions say that $R(x, wz)$ can be determined by first computing $\psi(x, w)$ and then determining $R(\psi(x, w), z)$. Since $|w| = \sigma(x) = O(\log |x|)$, the entire solution set $R(x)$ can be expressed as the disjoint union of a polynomial number of solution sets of the same relation on smaller instances.

Example 10.1. (*Spanning Trees*). We show here how the spanning trees in a graph can be coded in the language of this section. Let x represent a graph with n vertices. We suppose x is represented by its adjacency list which is indexed by $\ell = \lceil \log n \rceil$ bit integers. Spanning trees y can be represented as lists of $n - 1$ edges, each edge being a pair of vertex indices. Thus $\Sigma = \{0, 1\}$, and $R(x, y)$ if y is a spanning tree of x. Here, each y has $|y| = g(x) = 2\ell(n - 1)$. Let $\sigma(x) = 2\ell$ with the first $\sigma(x)$ characters of y representing the first edge in the list. For y of length $g(x)$, write $y = wz$, where $|w| = \sigma(x)$ and w represents one edge. Let $\psi(z, w)$ be the result of contracting the edge w in x (merging the vertices at the ends of w and erasing any resulting multiple edges). This yields a smaller graph: $|\psi(z, w)| \leq 2\ell(n-2) < |x|$. Note that $R(x, wz)$ if and only if $R(\psi(x, w), z)$.

That is, $y = wz$ is a spanning tree of x if and only if when we contract the edge w in x, z represents a spanning tree of the resulting $\psi(x, w)$.

Jerrum, Valiant, and Vazirani (1986) have given the following theorem which in essence says that for self reducible problems, efficient sampling yields efficient approximate counting. See Section 7 for the definition of polynomial relation.

THEOREM 10.2. *(Jerrum, Valiant, Vazirani). Let R be a self-reducible polynomial relation that is also in P. Suppose that we have an algorithm that*

- *takes input $x \in \Sigma^*$ and ϵ, $0 < \epsilon \leq 1$*
- *runs in time polynomial in $|x|$ and $\ln(1/\epsilon)$*
- *generates a random element $y \in R(x)$ whose distribution is within ratio $(1 \pm \epsilon)$ of the uniform distribution on $R(x)$.*

Then, given an x, ϵ, δ a random count c can be computed so that c approximates $|R(x)|$ within ratio $1 \pm \epsilon$, with probability at least $1 - \delta$. Moreover, this computation can be done in time polynomial in $|x|, 1/\epsilon$, and $\log(1/\delta)$.

Remark 10.3. Jerrum, Valiant and Vazirani (1986) show that the requirement that the relation R is in P can be dropped if a slightly different notion of near uniform sampling is used or if we work only with x for which $R(x)$ is nonempty.

We will now show that we can cast $\Sigma_{\mathbf{rc}}$ in a self-reducible form. For any table $F \in \Sigma_{\mathbf{rc}}$ and ordered pair $(k, \ell) \in [m] \times [n]$, let $[\Sigma_{\mathbf{rc}}]$ denote the set of all tables T in $\Sigma_{\mathbf{rc}}$ with

$$T_{ij} = F_{ij} \quad \text{and whenever} \quad i = k \quad \text{and} \quad j < \ell.$$

In other words, $[\Sigma_{\mathbf{rc}}|F; (k, \ell)]$ is the subset of $\Sigma_{\mathbf{rc}}$ tables whose entries match the table F in all positions *strictly* preceding (k, ℓ) in the lexicographic order. (We will use the symbols $\succ$ (and $\succeq$) for this order relation.) Notice that we have the following properties for any $F \in \Sigma_{\mathbf{rc}}$: $\Sigma_{\mathbf{rc}}$: (a) $F \in [\Sigma_{\mathbf{rc}}|F; (k, \ell)]$, (b) $[\Sigma_{\mathbf{rc}}|F; (1, 1)] = \Sigma_{\mathbf{rc}}$, and (c) for any $(k, \ell) \succ (m - 1, n - 1)$ we have $[\Sigma_{\mathbf{rc}}|F; (k, \ell)] = \{F\}$, since all remaining entries are then determined by the sum constraints.

If we use $F_{k\ell \leftarrow i}$ to denote the table obtained from F by setting $F_{k\ell} = i$, then we may write

$$[\Sigma_{\mathbf{rc}}|F; (k, \ell)] = \bigcup_{0 \leq i \leq N} [\Sigma_{\mathbf{rc}}|F_{k\ell \leftarrow i}; succ(k, \ell)),$$

where $succ(k, \ell)$ is the ordered pair that is the immediate successor of (k, ℓ) in the lexicographic order. Note that some of the sets on the right may be empty. This relation expresses the decomposition needed to show that

$[\Sigma_{\mathbf{rc}}|F;(k,\ell)]$ is polynomially self-reducible in the parameters in m, n, and N.

Now we will show that we can use a random walk to draw samples from $[\Sigma_{\mathbf{rc}}|F;(k,\ell)]$ as we used to draw sample from $\Sigma_{\mathbf{rc}}$. This walk is explained in section 10.1 above.

Algorithm 10.4. (Random walk on $[\Sigma_{\mathbf{rc}}|F;(k,\ell)]$). For a given k and ℓ, modify the basic walk of Section 10.1 so that it chooses only amongst values of i_1, i_2, j_1 and j_2 such that $(i_1, j_1) \succeq (k,\ell)$. That is, at each stage uniformly choose a pair of rows i_1 and i_2, $i_1 < i_2 \leq m$, and a pair of columns j_1 and j_2,

$$j_1 < j_2 \leq n, \qquad \text{such that} \quad (i_1, j_1) \succeq (k,\ell).$$

THEOREM 10.5. *The random walk generated by the Algorithm above and started on any $F \in \Sigma_{\mathbf{rc}}$ is ergodic and has uniform stationary distribution on $[\Sigma_{\mathbf{rc}}|F;(k,\ell)]$.*

Proof. We need to show that for each X and Y in $\Sigma_{\mathbf{rc}}$, there is a path between X and Y, using only possible steps of the walk. Note that since each step is reversible, a path from X to Y implies a symmetric one from Y to X.

We prove that there is a path joining X and Y by induction on a distance measure between the two tables X and Y. Define the distance $d(X,Y)$ between X and Y as $d(X,Y) = \sum_{i,j} |X_{ij} - Y_{ij}|$. Observe that $d(X,Y) = 0$ if and only if $X = Y$. Further note that, since the grand sums in the table are the same, this distance is always a multiple of two.

Let k be a nonnegative integer, and assume the following induction hypothesis: if $0 \leq d(X,Y) \leq 2k$ and if (i,j) is the lexicographically-first coordinate in which X and Y differ, then there is a path joining X and Y using only steps of the walk that do not involve coordinates lexicographically preceding (i,j). This is vacuously true for $k = 0$.

For the induction step, let X and Y be two elements of $\Sigma_{\mathbf{rc}}$ and suppose $d(X,Y) = 2(k+1)$, where (i,j) is the first coordinate in which they differ. We will show that either there is a move from X to X' where $d(X',Y) \leq 2k$ or there is a move from Y to Y' where $d(X,Y') \leq 2k$, where no coordinates preceding (i,j) are involved. The induction hypothesis will then imply that there is an entire path between X and Y.

In step 2 of the algorithm choose $i_1 = i$ and $j_1 = j$. Then there are two cases to consider.

Case 10.6. $(X_{i_1,j_1} < Y_{i_1,j_1})$. Then since each row and column of X has the same sum as in Y, we have

$$\exists j_2 \quad \text{such that} \quad X_{i_1,j_2} > Y_{i_1,j_2}$$
$$\exists i_2 \quad \text{such that} \quad X_{i_2,j_1} > Y_{i_2,j_1}.$$

We must have $i_1 < i_2$ and $j_1 < j_2$, since (i_1, j_1) was chosen as the lexicographically first position in which X and Y differ. Moreover, the entries X_{i_1,j_2} and X_{i_2,j_1} are both positive, since they are greater than their nonnegative counterparts in Y. This means, that letting $d = +1$, the move

$$X'_{i_1,j_1} = X_{i_1,j_1} + 1 \qquad X'_{i_1,j_2} = X_{i_1,j_2} - 1$$
$$X'_{i_2,j_1} = X_{i_2,j_1} - 1 \qquad X'_{i_2,j_2} = X_{i_2,j_2} + 1$$

yields an X' having nonnegative entries as well as sharing the same row and column sums as X.

By moving from X to X', the difference with respect to Y on least the three coordinates (i_1, j_1), (i_1, j_2), and (i_2, j_1) decreased by 1. The difference at (i_2, j_2) may have increased by 1, but the net change in all four coordinates must in any case be a decrease of at least 2. That is, $d(X', Y) \le d(X, Y) - 2$, so $d(X', Y') \le 2k$. Now by the induction hypothesis there is a path from X' to Y. Adding the step from X to X' completes the path from X to Y, without altering any coordinates lexicographically preceding (i, j) (in which X and Y already agree).

Case 10.7. $(X'_{i_1,j_1} > Y_{i_1,j_1})$. This case is entirely symmetric. Swapping the roles of X and Y, the same argument as in Case 10.6 shows that there is a move from Y to Y' with $d(X, Y') \le 2k$. Thus, by the induction hypothesis, there is a path between X and Y', and hence a path between X and Y via Y'.   □

Remark 10.8.   1.   To have a provably polynomial randomized algorithm it must be shown that the random walk above is rapidly mixing. This has been done for fixed dimensional problems in Diaconis and Saloff-Coste (1994). Here is a statement of their result. Let $P(x, y)$ be the transition matrix of the random walk described above. Here $x, y \in \Sigma_{\mathbf{rc}}$. Let U be the uniform distribution on $\Sigma_{\mathbf{rc}}$. Let γ be the diameter of $\Sigma_{\mathbf{rc}}$. This is the smallest n so $P^n(x, y) > 0$ for all $x, y \in \Sigma_{\mathbf{rc}}$.

THEOREM 10.9. *There are constants* a, b, α, β *such that*

$$\|P_x^k - U\| \le \alpha e^{-c} \quad \text{for} \quad k \ge ac\gamma^2 \quad \text{and all } x$$
$$\|P_x^k - U\| \ge \beta > 0 \quad \text{for} \quad k \le b\gamma^2 \quad \text{for some } x.$$

Here α, β, a, b *depend on the dimensions* m, n. *But not otherwise on* $\mathbf{r}, \mathbf{c}$.

This gives a randomized algorithm that runs in a polynomial number of steps in N for fixed m, n.

2.   Chung, Graham and Yau (1994) have announced much more sweeping results which are currently under close scrutiny. It seems likely that a definitive solution will be available.

3.   In practice, one uses much more vigorous random walks: one need not

move one each time and more complex patterns than the basic $\begin{matrix} + & - \\ - & + \end{matrix}$ can be used. Diaconis and Sturmfels contains further discussion and examples.

10.3. Convex sets for random generation. There is a somewhat different approach to random generation (and hence counting) which links into the healthy developments of computer science theory. Consider the set of all $m \times n$ arrays with *real* nonnegative entries, row sums $\mathbf{r}$ and column sums $\mathbf{c}$. This is a convex polyhedron containing $\Sigma_{\mathbf{rc}}$. There has been a great deal of work in the computer science literature on approximations to the volume of such convex polyhedra. Briefly, such problems are #-P complete but have polynomial time approximations using randomness. The already large literature on this subset is surveyed in Dyer and Frieze (1991). See Lovasz and Shimonovitz (1990) for recent results. It seems natural to try and adapt the ideas developed for the volume problem to the problem of counting $\Sigma_{\mathbf{rc}}$. Gangolli (1991) made an early application of these ideas to table enumeration. He needed to assume the row and column sums were fairly balanced.

Dyer, Kannan and Mount (1994) have recently made a breakthrough in this problem. Briefly, consider $\Sigma_{\mathbf{rc}}$ as a subset of lattice points in $(m - 1) \times (n - 1)$ dimensions. About each table, construct a parallelopiped with sides aligned to the lattice. Each such box is identified with a unique table. Let Σ be the convex hull of the union of these boxes. Points can be picked uniformly at random in Σ to good approximation using random walk. If the point is in one of the boxes, output the associated table as the choice. If the point is at the fringes, repeat.

A formal theorem requires a mild restriction on the row and column sums. Here is a simplified version of their results: suppose $r_i > n(n - 1)(m - 1)$ for each i and $c_j > m(n - 1)(n - 1)$ for each j. Then there is an algorithm for generating a random table with a distribution within ϵ of uniform invariation distance which runs in time bounded by a polynomial in $I, J, \max_{ij}(\log r_i, \log c_j)$ and $\log(1/\epsilon)$.

For m and n small, the restrictions on r_i and c_j allow tables of practical interest, for example, with $m = n = 4$. The restrictions are $r_i, c_j \geq 36$. The above is a simplified version of their work which actually is more general. The authors have done much more: they have implemented their algorithm and have it up and running. At present it produces about 1 table per second for tables like Table 6.1. The existence of an independent check of this sort has already been highly useful. It allowed us to pick up clear errors in some other algorithms.

Note Added in Proof. John Mount (1994) has made spectacular progress in deriving closed form expression for the number of tables. R.B. Holmes and L.K. Jones (1994), " *On the Uniform Generation of Two-Way tables with fixed margins and conditional volume test of Diaconis and Efron*", Technical Report, Dept of Mathematics, U-Mass-Lowell, offer a clever rejection algorithm for Monte Carlo sampling. Rates of convergence for the random walk of Section 10.2 are in Diaconis, P. and Saloff-Coste L., (1994), Random Walk on Contingency Tables with Fixed Row and Column Sums, Technical Report, Dept. of Mathematics, Harvard University.

Acknowledgement. We thank Nantel Bergeron, Fan Chung, Ira Gessel, I.J. Good, Ron Graham, Susan Holmes, Ravi Kannan, Brad Mann, John Mount, Richard Stanley, John Stembridge, and Bernd Sturmfels for their help with this project.

REFERENCES

[1] Baglivo, J. (1994) Forthcoming Book.

[2] Balmer, D. (1988) Recursive enumeration of $r \times c$ tables for exact likelihood evaluation, *Applied Statistics* **37**, 290-301.

[3] Békéssy, A., Békéssy, P., Komlós, J. (1972) Asymptotic enumeration of regular matrices, *Studia Sci. Math. Hung.* **7**, 343-353.

[4] Bender, E. (1974) The asymptotic number of non-negative integer matrices, with given row and column sums, *Discrete Math.* **10**, 217-223.

[5] Bishop, Y., Fienber, S., and Holland, P. (1975) *Discrete Multivariate Analysis*, MIT Press, Cambridge, MA.

[6] Bona, M. (1994) Bounds on the volume of the set of doubly stochastic matrices, Technical Report, Dept. of Mathematics, M.I.T.

[7] Boulton, D., and Wallace, C. (1973) Occupancy of a regular array, *Computing* **16**, 57-63.

[8] Chung, F., Graham, R., and Yau, S. (1994) On sampling in combinatorial structures, unpublished manuscript.

[9] Cormen, T., Leiserson, C., and Rivest, R. (1990) *Introduction to algorithms*, McGraw-Hill, New York.

[10] Curtis, C., and Reiner, I. (1962) *Representation Theory of Finite Groups and Associative Algebras*, Wiley Interscience, New York.

[11] Dahmen, W., and Micchelli, C. (1988) The number of solutions to linear Diophantine equations and multivariate splines, *Trans. Amer. Math. Soc.* **308**, 509-532.

[12] Danilov, V. (1978) The geometry of toric varieties, Russian Math. Surveys **33**, 97-154.

[13] Diaconis, P., and Efron, B. (1985) Testing for independence in a two-way table: new interpretations of the chi-square statistic (with discussion), *Ann. Statist.* **13**, 845-913.

[14] Diaconis, P. (1988) *Group representations in probability and statistics*, IMS Lecture Notes – Monograph Series, II. Institute of Mathematical Statistics, Hayward, CA.

[15] Diaconis, P. (1989) Spectral analysis for ranked data, *Ann. Statist.* **17**, 781-809.

[16] Diaconis, P., McGrath, M., and Pitman, J. (1993) Descents and random permutations, to appear, *Combinatorica*.

[17] Diaconis, P., and Saloff-Coste, L. (1993) Nash inequalities for Markov chains, Technical Report, Dept. of Statistics, Harvard University.

[18] Diaconis, P., and Sturmfels, B. (1993) Algebraic algorithms for sampling from conditional distributions, Technical Report, Dept. of Statistics, Stanford University.

[19] Dyer, A., and Frieze, A. (1991) Computing the volume of convex bodies: a case where randomness provably helps, in B. Bollobás, *Probabilistic Combinatorics and Its Applications*, Amer. Math. Soc., Providence, RI.

[20] Dyer, M., Kannan, R., and Mount, J. (1994) Unpublished manuscript.

[21] Everett, C., and Stein, P. (1971) The asymptotic number of integer stochastic matrices, *Discrete Math.* **1**, 55-72.

[22] Fisher, R.A. (1935) *The Design of Experiments*, Oliver and Boyd, London.

[23] Foulkes, H. (1976) Enumeration of permutations with prescribed up-down and inversion sequences, *Discrete Math.* **15**, 235-252.

[24] Foulkes, H. (1980) Eulerian numbers, Newcomb's problem and representations of symmetric groups, *Discrete Math.* **30**, 3-99.

[25] Fulton, W. (1993) *Introduction to Toric Varieties*, Princeton University Press, Princeton, NJ.

[26] Gail, M., and Mantel, N. (1977) Counting the number of $r \times c$ contingency tables with fixed margine, *Jour. Amer. Statist. Assoc.* **72**, 859-862.

[27] Gangolli, A. (1991) Convergence bounds for Markov chains and applications to sampling, Ph.D. Thesis, Computer Science Dept., Stanford University.

[28] Garey, M., and Johnson, D. (1979) *Computers and Intractability*, Freeman, San Francisco.

[29] Garsia, A. (1990) Combinatorics of the free Lie algebra and the symmetric group, *Analysis and etc.* (P.H. Rabinowitz and E. Zehnder, eds.), pp. 309-382. Academic Press, Boston.

[30] Garsia, A., and Reutenauer, C. (1989) A decomposition of Soloman's descent algebra, *Adv. Math.* **77**, 189-262.

[31] Gessel, I. (1990) Symmetric functions and *P*-recursiveness, *Jour. Comb. Th.* **A 53**, 257-285.

[32] Gessel, I., and Reutenauer, C. (1993) Counting permutations with given cycle structure and descent set, *Jour. Comb. Th.* **A 56**, 189-215.

[33] Good, I.J. (1976) On the application of symmetric Dirichlet distributions and their mixtures to contingency tables, *Ann. Statist.* **4**, 1159-1189.

[34] Good, I.J., and Crook, J. (1977) The enumeration of arrays and a generalization related to contingency tables, *Discrete Math.* **19**, 23-65.

[35] Goulden, I., Jackson, D. and Reilly, J. (1983) The Hammond series of a symmetric function and its application to *P*-recursiveness, *SIAM Jour., Alg. Discrete Methods* **4**, 179-183.

[36] Gusfield, D. (1988) A graph-theoretic approach to statistical data security, *SIAM Jour. Comp.* **17**, 552-571.

[37] Hancock, J. (1974) Remark on Algorithm 434, *Communications of the ACM* **18**, 117-119.

[38] Humphreys, J. (1990) *Reflection Groups and Coxeter Groups*, Cambridge University Press, Cambridge.

[39] Irving, R., and Jerrum, M. (1990) 3-D statistical data security problems, Technical Report, Dept. of Computing Science, University of Glasgow.

[40] Jackson, D., and Van Rees (1975) The enumeration of generalized double stochastic non-negative integer square matrices, *SIAM Jour. Comp.* **4**, 475-477.

[41] James, G.D., and Kerber, A. (1981) *The representation theory of the symmetric group*, Addison-Wesley, Reading, MA.

[42] Jerrum, M., and Sinclair, A. (1989) Approximating the permanent, *SIAM Jour. Comp.* **18**, 1149-1178.

[43] Jerrum, M., Valiant, L., and Vazirani, V. (1986) Random generation of combinatorial structures from a uniform distribution, *Theoretical Computer Science* **43**, 169-180.

[44] Jia, R. (1994) Symmetric magic squares and multivariate splines, Technical Report, Dept. of Mathematics, University of Alberta.

[45] Kerov, S.V., Kirillov, A.N., and Reshetikin, N.Y. (1986) Combinatorics, Bethe Ansatz, and representations of the symmetric group, translated from: *Zapiski Mauchnykh Seminarov Leningradskogo Otdeleniya Matematicheskogo Instituta im. V.A. Steklova An SSSR* **155**, 50–64.

[46] Knuth, D. (1970) Permutations, matrices, and generalized Young tableaux, *Pacific Jour.* **34**, 709-727.

[47] Knuth, D. (1973) *The Art of Computer Programming*, Vol. 3, Addison-Wesley, Reading, MA.

[48] Lovasz, L., and Simonovits, M. (1990). The mixing rate of Markov chains, an isoperimetric inequality, and computing the volume. Preprint 27, Hung. Acad. Sci.

[49] MacMahon, P. (1916) *Combinatorial Analysis*, Cambridge University Press, Cambridge.

[50] Macdonald, J. (1979) *Symmetric Functions and Hall Polynomial*, Clarendon, Press, Oxford.

[51] Mann, B. (1994) Unpublished manuscript.

[52] March, D. (1972) Exact probabilities for $r \times c$ contingency tables, *Communications of the ACM* **15**, 991-992.

[53] Mehta, C., and Patel, N. (1983) A network algorithm for performing Fisher's exact test in $r \times c$ contingency tables, *Jour. Amer. Statist. Assoc.* **78**, 427-434.

[54] Mehta, C., and Patel, N. (1992) Statexact.

[55] Mount, J. (1994) Ph.D. Thesis, Dept. of Computer Science, Carnegie Mellon University.

[56] O'Neil, P. (1969) Asymptotics and random matrices with row-sum and column-sum restrictions, *Bull. Amer. Math. Soc.* **75**, 1276-1282.

[57] Pagano, M., and Taylor-Halvorsen, K. (1981) An algorithm for finding the exact significance levels of $r \times c$ contingency tables, *Jour. Amer. Statist. Assoc.* **76**, 931-934.

[58] Sinclair, A. (1993) *Algorithms for random generalization and counting: a Markov chain approach*, Birkhäuser, Boston.

[59] Sinclair, A., and Jerrum, M. (1989) Approximate counting, uniform generation and rapidly mixing Markov chains, *Information and Computation* **82**, 93-133.

[60] Snee, R. (1974) Graphical display of two-way contingency tables, *Amer. Statistician* **38**, 9-12.

[61] Soloman, L. (1976) A Mackey formula in the group ring of a Coxeter group, *J. Alg.* **41**, 255-264.

[62] Stanley, R. (1973) Linear homogeneous Diophantine equations and magic labelings of graphs, *Duke Math. Jour.* **40**, 607-632.

[63] Stanley, R. (1986) *Enumerative Combinatorics*, Wadsworth, Monterey, CA.

[64] Stein, M.L., and Stein, P. (1970) Enumeration of stochastic matrices with integer elements, Los Alamos Scientific Laboratory Report, LA-4434.

[65] Valiant, L. (1979) The complexity of computing the permanent, *Theoretical Computer Sci.* **8**, 189-201.

THREE EXAMPLES OF MONTE-CARLO MARKOV CHAINS: AT THE INTERFACE BETWEEN STATISTICAL COMPUTING, COMPUTER SCIENCE, AND STATISTICAL MECHANICS

PERSI DIACONIS* AND SUSAN HOLMES[†]

Abstract. The revival of interest in Markov chains is based in part on their recent applicability in solving real world problems and in part on their ability to resolve issues in theoretical computer science. This paper presents three examples which are used to illustrate both parts: a Markov chain algorithm for estimating the tails of the bootstrap also illustrates the Jerrum-Sinclair theory of approximate counting. The Geyer-Thompson work on Monte-Carlo evaluation of maximum likelihood is compared with work on evaluation of the partition function. Finally, work of Diaconis-Sturmfels on conditional inference is complemented by the work of theoretical computer scientists on approximate computation of the volume of convex polyhedra.

Introduction

This paper presents three examples of what has come to be called the Markov chain simulation method. The examples blend together ideas from statistics, computer science, and statistical mechanics. The problems presented are set in statistical contexts of assessing variability, maximizing likelihoods, and carrying out goodness of fit tests. All of the examples involve reversible Markov chains on discrete sample spaces. In each case, the chains were actually run for a problem of real world interest. Each example is paired with a healthy theoretical development. As always, there are tensions and trade offs between practice and theory. This area brings them closer than usual.

1. The bootstrap and approximate counting

A. The bootstrap

Efron's bootstrap is a fundamental advance in statistical practice. It allows accurate estimation of variability without parametric assumptions. One begins with data $x_1, x_2, \cdots, x_n$ in a space $\mathcal{X}$. Let $T(x_1, x_2, \cdots, x_n)$ be a statistic of interest (e.g., a mean, median, correlation matrix,$\cdots$). We are interested in estimating the variability of T, assuming that x_i independent and identically chosen from an unknown distribution F on $\mathcal{X}$.

The bootstrap draws resample observations $x_1^*, x_2^*, \cdots, x_n^*$ from $\{x_1, x_2, \cdots, x_n\}$ with replacement and calculates $T(x_1^*, x_2^*, \cdots, x_n^*)$. Doing this repeatedly gives a set of values which can be proved to give a good indication of the distribution of T. A splendid up-to-date account of the bootstrap appears in Efron and Tibshirani (1993). Hall (1992) gives a solid

---

* Dept. of Mathematics, Harvard University, Cambridge, MA 02138.

† INRA, Unité de Biométrie, 2, Place Pierre Viala, 34060 Montpellier, France. (Visiting Stanford University).

theoretical development.

The present section explains a Monte Carlo method for deriving large deviations estimates of $P\{T \geq t\}$ when this probability is small. The idea is to run a Markov chain on the set of values $\{\underset{\sim}{x}^* = (x_1^*, \cdots, x_n^*) : T(\underset{\sim}{x}^*) \geq t\}$.

A step of the chain picks I, $1 \leq I \leq n$, uniformly at random and replaces x_I^* with a fresh value chosen uniformly from the original set of values $\{x_1, x_2, \cdots, x_n\}$. If the new sample vector satisfies $T(\underset{\sim}{x}) \geq t$, the change is made. Otherwise the chain stays at the previous sample vector.

This generates a reversible Markov chain on $\{\underset{\sim}{x} : T(\underset{\sim}{x}) \geq t\}$. It has a uniform stationary distribution. Assuming the chain is connected (see below) this gives an easy to implement method of sampling from the uniform distribution.

To estimate $P\{T \geq t\}$ we choose a grid $t_0 < t_1 < \cdots < t_\ell < t$ with t_0 chosen in the middle of the distribution of T and t_i chosen so that $P\{T \geq t_{i+1} | T \geq t_i\}$ is not too small. Here, all probabilities refer to the uniform distribution on the set of n-tuples chosen with repetition from $\{x_1, x_2, \cdots, x_n\}$.

With t_i chosen, first estimate $P\{T \geq t_0\}$ by ordinary Monte Carlo.

Then, estimate $P\{T \geq t_1 | T \geq t_0\}$ by running the Markov chain on $\{\underset{\sim}{x}^* : T(\underset{\sim}{x}^*) \geq t_0\}$

and counting what proportion of values satisfy the constraint $T \geq t_1$.

Continue, estimating $P\{T \geq t_2 | T \geq t_1\} \cdots P\{T \geq t | T \geq t_\ell\}$. Multiplying these estimates gives an estimate for $P\{T \geq t\}$:

$$\widehat{P}\{T \geq t\} = \widehat{P}\{T \geq t_0\} \widehat{P}\{T \geq t_1 | T \geq t_0\} \cdots \widehat{P}\{T \geq t | T \geq t_\ell\}.$$

Example 1.1. The following list of 10 pairs gives the average test score (LSAT) and grade point average (GPA) of 10 American law schools

LSAT	576	635	558	578	666	580	555	661	652	605
GPA	3.39	3.30	2.81	3.03	3.44	3.07	3.00	3.43	3.36	3.13

The scatter plot of these numbers in Figure 1.1 suggests a fair amount of association between LSAT and GPA. The correlation coefficient is $T = .81$. The bootstrap can be used to set confidence intervals for the true population correlation coefficient as suggested by Efron (1979). Figure 1.2 shows the result of 1000 repetitions of the basic bootstrap sampling procedure. This yields a 90% confidence interval [0.51, 0.99] for the population correlation coefficient. It provides a simple example of the use of the bootstrap. See Efron and Tibshirani (1993) for more detail.

We now turn to an example of the Monte Carlo Markov chain method. Suppose we want to estimate the proportion of the 10^{10} bootstrap samples with $T \geq .99$. We begin by choosing $t_0 = .9$ and estimating $P\{T > .9\} \doteq .4613$ based on 3000 Monte Carlo samples. Following this, take $t_i = .9i, 1 \leq i \leq 9$. The following table gives the Markov chain estimates of

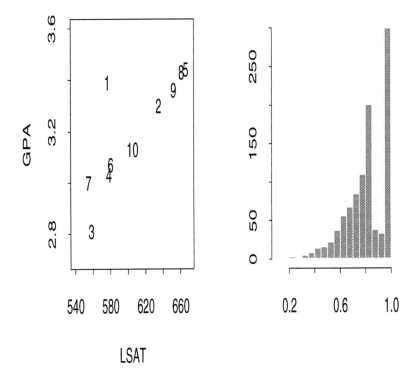

FIG. 1.1. *LSAT vs. GPA for 10 law schools*

FIG. 1.2. *1,000 bootstraps for correlations*

$P\{T \geq t_i | T \geq t_{i-1}\}$. These are each based on running the Markov chain described above 3000 steps.

t_i	.91	.92	.93	.94	.95	.96	.97	.98	.99
$P\{T \geq t_i | T \geq t_{i-1}\}$	.997	.972	.989	.961	.984	.918	.897	.766	.710

These result in the estimate $\widehat{P}\{T > .99\} = .177$. For this example, exact enumeration of the bootstrap distribution of T is possible (see Diaconis and Holmes (1994a)) and gives $P\{T \geq .99\} = .1769$.

Remark 1.2. (1) In general, t_i can be chosen sequentially as an approximate median of T on $\{\underset{\sim}{x}^* : T \geq t_{i-1}\}$. This can be estimated by a sequence of preliminary walks. In fact, all that is needed is that $P\{T \geq t_i | T \geq t_{i-1}\}$ is not exponentially small. In our experience, an intuitive choice of grid values does fine. We derive the optimal choice in Diaconis and Holmes (1994b).

(2) Rates of convergence for this type of Markov chain which can be used as a guide to sample size choice are starting to become available. See

Diaconis and Saloff-Coste (1993) and Sinclair (1993) for recent surveys.

(3) Confidence intervals for $\widehat{P}$ can be based on central limit and large deviations theorems for Markov chains. See Höglund (1974), Gilman (1993) and the references cited there.

(4) The Markov chain described above is one of a host of chains described in Diaconis and Holmes (1994b). For example, there is no need to change only one value at a time. Any set of values may be chosen. This can be important for insuring connectivity of the underlying chain.

(5) One advantage of changing only a single value: for many statistics (including the correlation coefficient) fast updating algorithms can be used to avoid complete recomputation. See Section 3 in Diaconis and Holmes (1994a) for further discussion and references.

B. Approximate counting

The algorithm described above is derived from 15 years of development in the computer science literature. They consider the problem of approximate counting. Let $\mathcal{X}$ be a finite set. The problem is to approximate $|\mathcal{X}|$. We are given the ability to choose at random from $\mathcal{X}$. Without further restrictions, the best one could do would be to wait for repeated values in the sample. This takes order $|\mathcal{X}|^{1/2}$ steps and is virtually useless for the large #-P complete problems of theoretical computer science.

Suppose further that there is a nested decreasing sequence of subsets $\mathcal{X} \supset \mathcal{X}_1 \supset \mathcal{X}_2 \cdots \supset \mathcal{X}_n$ with $k_i = |\mathcal{X}_i|/|\mathcal{X}_{i+1}|$ not too small, $k_0 = |\mathcal{X}|/|\mathcal{X}_1|$ and $|\mathcal{X}_n|$ small enough to be easily enumerated. We must also suppose the ability to sample uniformly from each $\mathcal{X}_i$ (at least approximately) one can then estimate $|\mathcal{X}_i|/|\mathcal{X}_{i+1}|$ by random sampling, providing an estimate denoted $\widehat{k}_i$ and then finally

$$\widehat{|\mathcal{X}|} = \widehat{k}_0 \widehat{k}_1 ... \widehat{k_{n-1}} \widehat{|\mathcal{X}_n|}$$

This is just the technique employed in Section A.

These ideas were introduced by Jerrum, Valiant, and Vazirani (1986) who showed that approximate counting and random generation are equivalent for self-reducible problems. Broder (1986) introduced the Markov chain aspect in his work on approximation of the permanent. If A is an $n \times n$ matrix $Per(A) = \sum_{\pi} \prod_{i=1}^{n} A_{i\pi(i)}$. The sum is over the symmetric group. If A is the adjacency matrix of a bipartite graph, $Per(A)$ counts the number of perfect matchings. Valiant (1979) has shown that evaluation of the number of matchings is #-P complete. Broder introduced a random walk on the space of matchings and proved that one could get good approximations to the number of perfect matchings in polynomial time if this walk was rapidly mixing by using just self-reducilibity. Rapid mixing of Broder's walk for dense graphs was proved by Jerrum and Sinclair (1989) who introduced several new ideas (conductance arguments and the use of paths). This necessarily brief history omits mention of several contributors (Alon, Mihail and others). It also omits a description of the sizable body of theory that has developed based on this work (Dyer, Frieze, Kannan, Lovasz, Simonovits and others). Fortunately, Sinclair (1993) gives a readable treatment of all of these issues.

One point is worth noting. The development in computer science took place in a theoretical context. The example in Section A may be close to the first implementation of these ideas. Evidently, applications present a field of further problems. One can only hope that the conversation continues.

2. Monte Carlo maximum likelihood and evaluation of partition functions

A. Maximum likelihood by Monte Carlo

In DNA fingerprinting problems one has samples of DNA from several individuals from which one would like to infer similarities. In one cleaned up version of the problem due to Geyer and Thompson (1992), the data consists of a binary matrix Y_{ij}, $1 \leq i \leq I$, $1 \leq j \leq J$, with each row representing an individual and the columns representing the lengths of DNA fragments. Geyer and Thompson used a model combining genetic insight with statistical mechanical simplicity. Let

$$U_i = \sum_j Y_{ij} = \# \text{ ones for } i^{\text{th}} \text{ individual}$$

$$V_j = \sum_i Y_{ij} = \# \text{ ones at } j^{\text{th}} \text{ level}$$

$$S_{ih} = \sum_j Y_{ij}Y_{hj} = \# \text{ common ones for } i^{\text{th}} \text{ and } h^{\text{th}} \text{ individual.}$$

The model postulates

$$(2.1) \quad P\{Y_{ij} \, l1 \leq i \leq I1 \leq j \leq J\} = c(\theta)^{-1} \exp\left\{\sum_i U_i \alpha_i + \sum_j V_j \beta_j + \sum_{i<k} S_{ih}\gamma_{ih}\right\}$$

Here, $\theta = \{\alpha_i, \beta_j, \gamma_{ih}\}$ has dimension $I + J + J(J-1)/2$ and $c(\theta)$ is a normalising factor. Genetic considerations force the constraint $\gamma_{ih} \geq 0$.

One problem now is, given a realization Y_{ij}, $1 \leq i \leq I$, $1 \leq j \leq J$, find the vector of parameters $\hat{\theta}$ that maximizes the likelihood. In one of their examples, $I = 79$, $J = 32$. This leads to 607 parameters to be estimated. The practical version of the problem has additional complications: some binary values were missing and extensive prior knowledge of relatedness was applicable.

One contribution of the Geyer-Thompson work is a novel method for maximizing likelihoods. This is generally applicable and will be explained now. We begin by defining a general exponential family. Let $\mathcal{X}$ be a set and $\mu(dx)$ a measure defined on a suitable class of subsets of $\mathcal{X}$. Let $T : \mathcal{X} \to \mathbb{R}^d$. The exponential family through T, μ is the family of probability measures

$$P_\theta(dx) = c(\theta)^{-1} e^{T(x)\cdot\theta} \mu(dx), \quad \theta \in \Theta.$$

In (2.2) $c(\theta)$ is a normalizing constant and Θ is a prespecified subset of $\mathbb{R}^d$ such that

$$c(\theta) = \int e^{T(x)\cdot\theta}\mu(dx) < \infty \text{ for } \theta \in \Theta.$$

For fixed x, the maximum likelihood estimator (MLE) $\hat{\theta} = \theta(x)$ is the value minimizing $-\log P_\theta(dx) = -\log(c(\theta)) - T(x)\cdot\theta$. Thus computation of the MLE requires knowledge of the normalizing constant. Typically, this is not available in closed form.

Geyer and Thompson (1992) suggest a Monte Carlo approach for this: fix a value $\theta_0 \in \Theta$ and write

$$
\begin{aligned}
c(\theta) &= \int e^{T(x)\cdot\theta}\mu(dx) = c(\theta_0) \int e^{T(x)\cdot(\theta-\theta_0)}\frac{e^{T(x)\cdot\theta_0}}{c(\theta_0)}\mu(dx) \\
&= c(\theta_0) \int e^{T(x)(\theta-\theta_0)} P_{\theta_0}(dx).
\end{aligned}
$$

Run a Markov chain $X_0, X_1, X_2, \cdots$ with stationary distribution P_{θ_0}. Then,

$$d_n = \frac{1}{n}\sum_{n=1}^{n} e^{T(x_i)(\theta-\theta_0)} \longrightarrow c(\theta)/c(\theta_0)$$

for large n. So

$$-\hat{\ell}_n = -\log d_n - T(x)\cdot\theta \longrightarrow \log c(\theta_0) - \log c(\theta) - T(x)\cdot\theta.$$

So the minimizer of $\hat{\ell}_n$ approximates $\hat{\theta}$. Practically, θ_0 should be well chosen else the chain will take a long time to converge. The following sequential scheme is used: after a few iterations with θ_0, replace θ_0 by $\hat{\theta}$ and iterate this a few times.

In the DNA fingerprinting example, Geyer and Thompson carried out such an analysis for the model (2.1). They used the standard Metropolis algorithm changing single sites of $\{Y_{ij}\}$ and thinning down as usual to run a Markov chain with stationary distribution P_{θ_0}. The estimated parameters were used to compute correlations between observable quantities. The resulting correlation matrices were used to cluster the data. The results made sense in the context of the original problem and seemed useful. We refer to Geyer and Thompson (1992,1993) for examples and further details.

B. Computing partition functions

In statistical mechanics, the normalizing constant $c(\theta)$ is called the partition function. There has been a good deal of careful mathematical work which allows one to prove that exact evaluation or even good approximation of $c(\theta)$ *even at a fixed value of* θ is intractable in most cases of interest. Welsh (1993) and Jerrum-Sinclair (1991) review this work. These last authors also give an approximation scheme for the widely-studied case of the ferro-magnetic Ising model which uses randomness.

The Ising model is a standard exponential family for variables $Y_i \in \{-1, +1\}$, $1 \leq i \leq n$, with

$$P_\theta(Y_i; 1 \leq i \leq n) = c(\theta)^{-1} e^{\Sigma_{i,j} \gamma_{ij} Y_i Y_j - \beta \Sigma_i Y_i}.$$

Here γ_{ij} are constrained to be zero unless (i, j) is in E, the set of edges of a prespecified graph (often a rectangular lattice). If γ_{ij} are non-negative, one is said to be in the ferromagnetic case: configurations with high probability tend to be all the same (all ones or all minus ones). Jerrum and Sinclair (1991) gave a novel algorithm for estimating $c(\theta)$ in the ferromagnetic case. Their algorithm approximates $c(\theta)$ to relative error $(1 \pm \epsilon)$. They prove that their algorithm works with probability $1 - \delta$ and requires only a polynomial number of steps in $|E|, n, \log \frac{1}{\delta}$ and $1/\epsilon$, uniformly in θ.

It is worth remarking that any use of the Metropolis algorithm for estimating $c(\theta)$ is doomed to fail. Indeed, in dimension two or higher, Ising models have phase transitions which means that there are (at least) two high energy wells and the chain will spend an exponential time in one of them and so fail to mix rapidly.

Jerrum and Sinclair overcome this difficulty by re-expressing the partition function in a radically different way derived by physicists.

$$c(\theta) = A \sum_{X < E} W(X)$$

where the sum is over all subgraphs X of the underlying graph having all vertices of odd degree, A is a simple, easy to compute function of θ, and

$$W(X) = \mu^{|X|} \cdot \prod_{(i,j) \in X} \lambda_{ij}$$

with $\lambda_{ij} = \tanh(\gamma_{ij})$, $\mu = \tanh(\beta)$ and $|X|$ the number of vertices in X. The weights $W(X)$ are all positive and this allows Jerrum and Sinclair to run a Markov chain on the set of subgraphs X with stationary distribution proportional to $W(X)$. This is not the end of the story; there is a clever new idea that allows a good estimate $\widehat{c}(\theta)$ of $c(\theta)$ from this chain. For now, we will stop and state their main result as

THEOREM 2.1. *(Jerrum and Sinclair) Let E be a graph with $|E|$ edges and n vertices. For all $\gamma_{ij} \geq 0$ and β real.*

$$P\{1 - \epsilon \leq \left| \frac{\widehat{c}(\theta)}{c(\theta)} \right| \leq 1 + \epsilon\} > 1 - \delta.$$

Here $\widehat{c}$ *is an estimator of c based on*

$$O(\epsilon^{-2} |E|^2 \mu^{-4} \{\log(\delta^{-1}) + |E|\}) \text{ operations.}$$

The implicit constants are uniform in θ.

Remark 2.2. We find a comparison between the Geyer-Thompson and Jerrum-Sinclair work instructive. First, although problems and techniques are clearly related, these authors worked without knowledge of each other or associated literature. Geyer-Thompson tried their ideas out in several examples and proved nothing beyond convergence. Jerrum and Sinclair's work strongly suggests that what Geyer-Thompson were trying to do was impossible!

Jerrum and Sinclair proved convergence in polynomial time, but did not implement it. In a personal communication they suggest that although slow, their algorithm should be implementable, perhaps running with the same efficiency as the volume algorithm studied in the next section.

Their work suggests that there is much more proving and comparing to be done before the algorithms of Geyer-Thompson are accepted: there are versions of the Ising model where the partition function is known or accurately computable (e.g., for planar graphs). At the least the algorithms should be tried out here. In the other direction, we would like to encourage computer scientists to get implementable versions of their algorithms. It often leads to fascinating insights and seems like a simple embarrassment which the rest of the world has really noticed.

3. Contingency tables and approximate volumes

A. Random walks on contingency tables

A contingency table is an $I \times J$ array of non-negative integers. These arise in statistical analysis of cross-classified data (e.g., a 4×7 table might represent a classification of students by class (freshman, sophomore, junior, senior) amd seven categories of extra-curricular activity. We will work with tables having fixed row sums $(r_1, r_2, \cdots, r_I)$ and fixed column sums $(c_1, c_2, \cdots, c_J)$. Thus define

$$\mathcal{X}(\underset{\sim}{r}, \underset{\sim}{c}) = \{x_{ij} : \sum_j x_{ij} = r_i, \ \sum_i x_{ij} = c_j, \ 1 \leq i \leq I, \ 1 \leq j \leq J\}.$$

The problem considered is choosing a table uniformly at random from $\mathcal{X}(\underset{\sim}{r}, \underset{\sim}{c})$.

This problem arises in statistical work and in many areas of combinatorics. A survey of applications is given in the article of Diaconis and Gangolli in this volume. For even moderate size tables (e.g., 4×7) the enumeration problem gets wildly out of hand. The following simple random walk algorithm gives a satisfactory method of proceeding.

Pick a pair of rows (i, i') at random and a pair of columns (j, j') at random. These intersect in 4 entries:

$$
\begin{array}{cc}
 & j \quad j' \\
\begin{array}{c} i \\ i' \end{array} & \begin{pmatrix} \cdot & \cdot \\ \cdot & \cdot \end{pmatrix}
\end{array}
$$

These entries are changed by adding and subtracting 1 from the 4 entries in the pattern $\begin{smallmatrix} + & - \\ - & + \end{smallmatrix}$ or $\begin{smallmatrix} - & + \\ + & - \end{smallmatrix}$ with probability $1/2$. This doesn't change the row or column sums. If it forces negative entries, the step is not taken (the walk stays where it was last). It is clear that this walk is symmetric: the chance of going from x to y in one step is the same as the chance of going from y to x: one must pick the same rows and columns and the opposite pattern. It is easy to see that this walk is connected. A connected symmetric walk on a finite set converges to the uniform distribution (there is some holding probability so there are no periodicity problems).

There has been extensive practical work based on this algorithm and many variations (e.g., one needn't move one each time and more complicated patterns can be used). These algorithms appear to converge rapidly and seem to make a previously intractable problem easy.

Here is an example: in testing for statistical independence of row and column effects one often uses the chi-squared statistic

$$T(x) = \sum_{i,j} \frac{\{x_{ij} - r_i c_j/n\}^2}{r_i c_j/n}.$$

Here $n = r_1 + \cdots + r_I = c_1 + \cdots + c_J$. Diaconis and Efron (1986) wanted to know the distribution on $\mathcal{X}(\underset{\sim}{r}, \underset{\sim}{c})$. As an example, consider the 5×3 table

5	2	3	10
50	7	5	62
3	6	4	13
5	3	3	11
2	7	30	39
65	25	45	135

This table has $T(x) = 72.18$. Gangolli (1991) ran the Markov chain algorithm and recorded the values of $T(x)$ as the chain progressed. A histogram of these values is shown in Figure 3.1 below , it is to be noted that these computations took about 1/30th of the time of the exhaustive method (20 minutes) . In practice, one would use this histogram to estimate the proportion of tables with $T(x) \leq 72.18$ which is about .76086.

Gangolli (1991) also gave a complete enumeration: there are 239,382,173 tables in $\mathcal{X}(\underset{\sim}{r}, \underset{\sim}{c})$, this enumeration took 15 hours and 21 minutes of real time . Based on these, a complete enumeration of the distribution of T is shown in Figure 3.2. The simulation and the truth match closely.

These contingency table problems arise in many variations and extensions (e.g., 3-dimensional arrays). A calculus for deriving basic moves that uses algebraic geometry is given by Diaconis and Sturmfels (1993) who also discuss the statistical literature and competing algorithms. Diaconis, Graham and Sturmfels (1993) give further extensions. A rigorous analysis of the running times of the suggested walks that applies to small tables (the case of most interest in statistical work) appears in Diaconis and Saloff-Coste (1993). Chung, Graham, and Yau (1994) have announced more powerful results. Gangolli (1991) has developed tempting conjectures about "the right answer" for running times. Mann (1994) gives exact formulae for $2 \times J$ and $3 \times J$ tables which permit checks of validity.

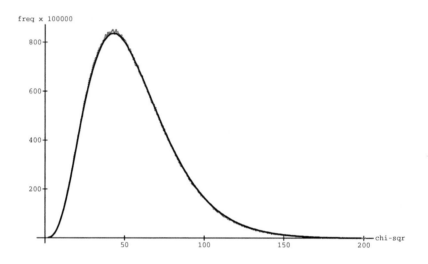

FIG. 3.1. *The exact distribution of* $T(x)$ *(in black)*

B. Approximating the volume

The work described above has close connections to a very healthy development in theoretical computer science. This asks for ways of computing the volume of a convex polyhedra in $\mathbb{R}^d$. There is a clear intuitive connection; the set $\mathcal{X}(\underset{\sim}{r}, \underset{\sim}{c})$ consists of the integer points in the convex polyhedron of all non-negative real arrays with a given set of row and column sums.

The volume problem is clearly basic. The problem is also hard: technically #-P complete. In fact, if use of randomness is forbidden, it can be proved that it is hard to get an approximate answer to within a factor of 2. Careful description and references are in the readable survey paper of Dyer and Frieze (1991).

Recent work shows that it is possible to get good approximations to the volume accurate to within a factor of $1 \pm \epsilon$, in a polynomial number of operations. Here, the parameters governing the problem may be taken as N – the number of hyperplanes specifying the polyhedron. The current best result, due to Lovasz-Kannan-Shimonovitz (1994) is a polynomial of degree 5 in N and $\log(1/\epsilon)$. The procedure uses a rapidly mixing Markov

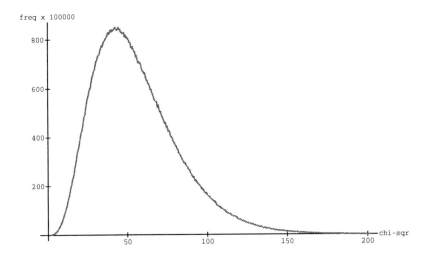

FIG. 3.2. *A Monte-Carlo approximation to Figure 3.1*

chain to sample from the polyhedron. As this result is of independent interest in statistical applications, we will spell it out.

Let C be a convex region in $\mathbb{R}^d$. Let $f(x)$ be a log concave probability density on C. The problem is to sample from $f(x)$. As an example, C might be the orthant

$$C = \{x_1 < x_2 < \cdots < x_d\}$$

and $f(x)$ might be the d-dimensional normal density (with mean vector μ) restricted to C. This is the problem of simulating normal vectors with a given order structure.

A simple algorithm runs as follows: starting from $x \in C$, pick a point on a small ball of radius δ uniformly. Now use the Metropolis algorithm to thin down this uniform walk to have density $f(x)$. This produces a reversible Markov chain with stationaary distribution $f(x)$ on C. The actual algorithm has some further ideas: The body C is first rounded by an affine transformation and there is some art in choosing a suitable δ. The proof of rapid mixing uses conductance which is bounded by a version of the Payne-Weinberger theorem of differential geometry. This coming together of different fields: statistics, convex geometry, differential geometry, linear programming and the theory of algorithms seems truly exciting.

At the workshop, Ravi Kannan some spectacular work with Dyer and Mount. They have found a way to adapt the continuous convex set problem to generate contingency tables! Briefly, their idea is this: consider the tables as lattice points in a convex polyhedra of dimension $(I - 1)(J - 1)$. Each table is in a small "box" of the same size. Adding such boxes for tables near the edge makes a non-convex figure. They take the convex hull of this. Now, the continuous algorithm is used to sample from the uniform distribution on the augmented figure. For each sample point, one associates the table labeling the box containing the point.

Kannan and Mount prove that this generates approximately uniform tables in a polynomial number of steps.

The Kannan and Mount work requires a mild restriction on the row and column sums to guarantee success. Here is the essence of their results: Suppose $r_i > J(J-1)(I-1)$ for each i and $c_j > I(I-1)(J-1)$ for each j. Then, there is an algorithm for generating a random table distribution within ϵ of uniform in variation distance which runs in time bounded by a polynomial in I, J, $\max_{i,j}(\log r_i, c_j)$ and $\log(1/\epsilon)$.

For I and J small, the restrictions on r_i and c_j allow tables of practical interest. For example with $I = J = 4$ the restrictions are $r_i, c_j \geq 36$. The above is a simplified version of their work which is actually more general.

Moreover, they have programmed versions of their algorithm which at the time of this writing produces one "clean" table per second. It has given an independent check on other procedures, which revealed an embarrassing error had been made. One can hope that it can be adopted for other Monte Carlo procedures proposed by Diaconis and Sturmfels (1993).

4. Related literature

The use of Markov chains is in a phase of seemingly exponential growth. We have pointed to some of the pieces above. A completely independent development is occuring in the area of image processing. The recent survey volume by Barone, Frigessi and Piccioni (1992) gives a good set of pointers to this literature. There is an equally healthy development in the language of statistical computing. Volume 55, *no.*1 of the Journal of the Royal Statistical Society has several surveys and discussions by leading workers in this field. Finally we mention that workers in the computational side of statistical mechanics have not stopped with the 'Metropolis' algorithm! There has been a steady development over the years. One way to access this is to browse through the last few years of the *Journal of Statistical Mechanics*.

The work described leaves a rich legacy of problems for probablists and statisticians. Usually, no hint of rates of convergence are available (aside from essentially meaningless statements about "exponential convergence" with unspecified constants in and in front of the exponent). The statistical problems involved when making Markov chain runs also seem wide open. For instance, should one use all the data afterr an inital inhibiting time, or even after the initial period only use widely spaced instances. The first case provides better accuracy, but the correlation involved is difficult to estimate. The latter pays by lack of efficiency for providing a standard estimate forr the correlation structure. Another aspect yet to be addressed is how to take into account any a priori knowledge.

REFERENCES

Barone,P., Frigessi A. and Piccioni M. (editors), 1992, Stochastic Models , Statistical Methods, and Algorithms in Image Analysis, Lecture Notes in Statistics, Springer–Verlag.

Broder, A. (1986). How Hard is it to Marry at Random? (On the approximation of the permanent.) *Proc. 18th ACM Symp. Th. Comp.*, 50–58. Erratum in *Proc. 20th ACM Symp. Th. Comp.*, 551.

Chung, F., Graham, R., and Yau, S.T. (1994). Unpublished manuscript.

Diaconis, P. and Efron, B. (1986). Testing for independence in a two-way table: new interpretations of the chi-square statistic (with discussion). *Ann. Statist.* **13**, 845–905.

Diaconis, P. and Gangolli, A. (1994). The number of arrays with given row and column sums. In this volume.

Diaconis, P., Graham, R.L., and Sturmfels, B. (1994). Primitive Partition Identities. Technical Report No. 9, Dept. of Statistics, Stanford University.

Diaconis, P. and Holmes, S. (1994a). Gray codes for randomization procedures. Technical Report No. 10, Dept. of Statistics, Stanford University.

Diaconis, P. and Holmes, S. (1994b). Random walks for bootstrap tails. Technical Report, Dept. of Mathematics, Harvard University.

Diaconis, P. and Saloff-Coste, L. (1993). Comparison theorems for reversible Markov chains. *Ann. Appl. Prob.* **3**, 696–730.

Diaconis, P. and Sturmfels, B. (1993). Algebraic algorithms for sampling from conditional distributions. Technical Report, Dept. of Mathematics, Harvard University.

Dyer, M. and Frieze, A. (1991). Computing the volume of convex bodies: a case where randomness proveably helps. *Probabilistic Combinatorics and its Applications*, B. Bollobàs, (ed.), *Proc. Symp. Appl. Math.* **44**, 123–170, Amer. Math. Soc., Providence.

Efron, B. (1979). Bootstrap methods: another look at the jackknife. *Ann. Statist.*, **7**, 1–26.

Efron, B. and Tibshirani, R. (1993). *An Introduction to the Bootstrap.* Chapman and Hall.

Gangolli, A. (1991). Convergence bounds for Markov chains and applications to sampling. Ph.D. Dissertation, Dept. of Computer Science, Stanford University.

Geyer, C. and Thompson, E. (1992). Constrained Monte Carlo maximum likelihood for dependent data. *Jour. Roy. Statist. Soc.* B **54**, 657–699.

Geyer, C. and Thompson, E. (1993). Analysis of relatedness in the California condors, from DNA fingerprints. *Mol. Biol. Evol.* **10**, 571-589.

Gillman, D. (1993). A Chernoff bound for random walks on expander graphs. Preprint.

Hall, P. (1992). *The Bootstrap and Edgeworth Expansions*, Springer–Verlag.

Höglund, T. (1974). Central limit theorems and statistical inference for finite Markov chains. *Zeit. Wahr. Verw. Gebeits.* **29**, 123–151.

Jerrum, M. and Sinclair, A. (1989). Approximating the permanent. *Siam J. Comput.* **18**, 1149–1178.

Jerrum, M. and Sinclair, A. (1993). Polynomial time approximation algorithms for the Ising model, *Siam J. Comp.*, **22**, 1087-1116.

Jerrum, M., Valiant, L. and Vazirani, V. (1986). Random generation of combinatorial structures from a uniform distribution, *Theor. Comput. Sci.* **43**, 169–188.

Kannan, R., Lovasz, L. and Simonovitz, M. (1994). Unpublished manuscript.

Kannan, R. and Mount, J. (1994). Unpublished manuscript.

Mann, B. (1994). Some formulae for enumeration of contingency tables. Technical Report, Harvard University.

Sinclair, A. (1993). *Algorithms for random generation and counting*, Birkhäuser, Boston.

Valiant, L. (1979). The complexity of computing the permanent. *Theor. Comput. Sci.* **8**, 189–201.

Welsh, D. (1993). *Complexity : Knots, colourings and counting*, Cambridge University Press.

THE MOVE-TO-FRONT RULE FOR SELF-ORGANIZING LISTS WITH MARKOV DEPENDENT REQUESTS*

ROBERT P. DOBROW† AND JAMES ALLEN FILL‡

Abstract. We consider the move-to-front self-organizing linear search heuristic where the sequence of record requests is a Markov chain. Formulas are derived for the transition probabilities and stationary distribution of the permutation chain. The spectral structure of the chain is presented explicitly. Bounds on the discrepancy from stationarity for the permutation chain are computed in terms of the corresponding discrepancy for the request chain, both for separation and for total variation distance.

AMS(MOS) subject classifications. Primary 60J10; secondary 68P10, 68P05.

Key words. Markov chains, self-organizing search, move-to-front rule, convergence to stationarity, separation, total variation distance, coupling.

1. Introduction and summary. A collection of n records is arranged in a sequential list. Associated with the ith record is a weight r_i measuring the long-run frequency of its use. We assume that each $r_i > 0$ and normalize so that $\sum r_i = 1$. At each unit of time, item i is removed from the list with probability r_i and replaced at the front of the list. This gives a Markov chain on the permutation group S_n.

If we assume that items are requested independently of all other requests, this model for dynamically organizing a sequential file is known as the move-to-front (MTF) heuristic and has been studied extensively for over 20 years. Background references include Rivest (1976), Bitner (1979), Hendricks (1989), Diaconis (1993), and Fill (1995). In the case when all the weights are equal the model corresponds to a card-shuffling scheme known as the random-1-to-top shuffle; see Diaconis et al. (1992) for a thorough analysis in this case.

One objection to this model is that in practice record requests tend to exhibit "locality of reference." That is, frequencies of access over the short run may differ quite substantially from those over the long run. Knuth (1973) and Bentley and McGeoch (1985), among others, have noted that MTF tends to work even better in practice than predicted from the i.i.d. model. Knuth cites computational experiments involving compiler symbol tables and notes that typically "successive searches are not independent (small groups of keys tend to occur in bunches)."

Konnecker and Varol (1981) proposed modeling the request sequence along Markovian or autocorrelative lines. Lam et al. (1984) formally set up

---

* Research for both authors supported by NSF grant DMS–93–11367.

† NIST, Administration Building, A337, Gaithersburg, MD 20899–0001. E-mail Address: dobrow@cam.nist.gov

‡ The Department of Mathematical Sciences, Johns Hopkins University, 34th and Charles Streets, Baltimore, MD 21218-2689. E-mail Address: jimfill@jhuvm.hcf.jhu.edu

the Markovian model for self-organizing search and obtained a formula for the asymptotic average cost of searching for a record (i.e., for stationary expected search cost). Phatarfod and Dyte (1993) have derived the eigenvalues of the transition matrix for this *Markov move-to-front* (MMTF) model.

In this paper we assume the Markovian model. We call the Markov chain corresponding to the sequence of record requests the *request chain.* We derive explicit formulas for the transition probabilities and stationary distribution of Markov move-to-front. We also study convergence to stationarity for MMTF and obtain bounds on separation and total variation distance of the MMTF chain from its stationary distribution in terms of the discrepancy from stationarity for the request chain.

After setting notation and addressing preliminary issues we discuss models for the request chain in Section 2. In Section 3, the k-step transition probabilities for MMTF are derived. In Section 4, we give three formulas for the stationary distribution. One has a direct probabilistic interpretation; the other two are more suited for numerical calculation in terms of the request matrix or its time reversal. In Section 5 we give the spectral structure of the MMTF chain. In the final two sections we analyze the speed of convergence of MMTF to its stationary distribution. In Section 6 we treat separation and obtain bounds in terms of the separation of the request chain. In Section 7 we treat total variation distance. In brief, our approach there is as follows. Variation distance is bounded above by the tails of any coupling time. We couple two copies of the MMTF chain by first coupling the corresponding request chains and then using the standard coupling for MTF, namely, wait until all but one of the records has been requested at least once.

2. Preliminaries and discussion of models. Let $\sigma \in S_n$ represent an ordered list of records, with σ_k denoting the record at the kth position in the list. Let $r_1, \ldots, r_n$ be a sequence of probabilities (weights), with the interpretation that record i is requested with long-run frequency r_i. We assume that all the probabilities are strictly positive. Let $[n] := \{1, \ldots, n\}$.

Let R be the $n \times n$ transition matrix for the request chain. Thus $R(i, j)$ is the probability of accessing record j given that the previous request was for record i. In the case of independent requests the rows of R are identical and equal to $(r_1, \ldots, r_n)$. We will denote such a request matrix by R_0 and refer to MMTF with such a request matrix as the *i.i.d. case* or as MTF.

Several authors (e.g., Lam et al. (1984) and Kapoor and Reingold (1991)) have considered the model

$$R := (1 - \alpha)R_0 + \alpha I_n$$

for the request chain R, where I_n is the $n \times n$ identity matrix and $\alpha \in [0, 1]$. The results of this paper can be easily applied to this case using the fact that

$$Q^k = \sum_{j=0}^{k} \binom{k}{j} (1 - \alpha)^j \alpha^{k-j} Q_0^j, \quad k \geq 0,$$

where Q_0 is the transition matrix for MTF and Q is the transition matrix for MMTF.

A generalization of this model which seems to capture locality of reference reasonably well is a mixture of the i.i.d. chain and a birth-and-death chain, that is,

$$(2.1) \qquad\qquad R := (1 - \alpha)R_0 + \alpha B,$$

where B is a birth-and-death transition matrix. Unfortunately, we do not know much in the way of neat formulas for this case. The analysis leads to quite difficult problems involving taboo probabilities and covering times for Markov chains. We shall, however, treat the extreme case when $\alpha = 1$ in (2.1), that is, the *birth-and-death case* $R = B$. This example may not be so realistic in that it exhibits *too much* locality of reference. However, it does provide an interesting and non-trivial request chain for which fairly complete results can be obtained.

Another generalization of the model $R = (1 - \alpha)R_0 + \alpha I_n$, allowing the holding probability α to vary with the record, is treated in Dobrow (1994) and in Rodrigues (1993).

There are several ways to model MMTF. Lam et al. (1984) model the chain on the state space $S_n \times [n]$, where (σ, i) denotes (present configuration, next request). This approach has the advantage of being able to handle other self-organizing schemes besides move-to-front. It is clear, as Phatarfod and Dyte (1993) point out, that the state space can also be specified as (present configuration, last request). Since for MMTF the configuration itself incorporates information about the last request—the last request can be read from σ as σ_1—we can and do take S_n as the state space for MMTF.

Let Q be the transition matrix for MMTF. Then

$$(2.2) \quad Q(\pi, \sigma) = \begin{cases} R(\pi_1, \sigma_1), & \text{if } \sigma = \pi \circ (k \cdots 1), \text{ where } k = \pi^{-1}(\sigma_1), \\ 0, & \text{otherwise.} \end{cases}$$

In all that follows we require that the R-chain be ergodic, which implies that its stationary distribution is unique and strictly positive. Note that this does *not* imply that the permutation chain is ergodic; see Section 4 for further discussion.

3. Transition probabilities. Before proceeding to the main result of this section (Theorem 3.1) we establish some terminology and notation. Say that a permutation σ has a *descent* at position i if $\sigma_i > \sigma_{i+1}$. Denote by $L(\sigma)$ the last descent position for σ, with $L(\text{id}) := 0$, where id denotes the identity permutation. Note that $L(\pi^{-1}\sigma)$ is the minimum number of entries of π that must be moved to the front to obtain σ.

In what follows, the time-reversal of the R chain will arise naturally. Let $\tilde{R}$ be the transition matrix of the time-reversed chain; that is, $\tilde{R}(x,y) := r_y R(y,x)/r_x$, where $\mathbf{r} = (r_1, \ldots, r_n)$ is the stationary distribution of the R chain.

For a vector or permutation σ of length at least m, we write $\sigma_{\to m}$ to denote the m-element vector consisting of the first m elements of σ. Notation such as $y_{t \to m}$ is shortand for $(y_{t_1}, \ldots, y_{t_m})$. We write $\sigma[m]$ to denote the unordered set $\{\sigma_1, \ldots, \sigma_m\}$. For a matrix A and set S, let A_S denote the principal submatrix of A determined by the rows and columns identified by S.

Let $\tilde{X} = (\tilde{X}_k)_{k \geq 0}$ be a Markov chain with transition matrix $\tilde{R}$. We use the notation $P_x(\cdot)$ for conditional probability given that $\tilde{X}_0 = x$ and $P_{\mathbf{r}}(\cdot)$ for probability with respect to the $\tilde{X}$-chain started in its stationary distribution $\mathbf{r}$.

We next define the *partial cover time* $\tilde{C}_m$, the first time at which $\tilde{X}$ has visited m distinct states. Formally, we can define $\tilde{C}_m$ as follows:

$$\tilde{C}_m := \inf\{k \geq 0 : |\tilde{X}[k]| = m\},$$

recalling that $\tilde{X}[k]$ denotes the unordered set $\{\tilde{X}_1, \ldots, \tilde{X}_k\}$. In particular, $\tilde{C}_1 = 0$, $\tilde{C}_n = \tilde{C}$ is the usual cover time for $\tilde{X}$, and $\tilde{C}_m = \infty$ for $m > n$.

A key observation for analyzing the MTF chain, utilized by Fill (1995) in obtaining an exact formula for the k-step transition probabilities, is that one can read off the sequence of last requests of distinct records from the order of the list. That is, if $\sigma \in S_n$ is the final order of the list, then σ_1 must have been requested last, σ_2 is the penultimate distinct record to have been requested, etc. We will make use of this observation in the proof of the following theorem.

THEOREM 3.1. *Let $\pi, \sigma \in S_n$. Under MMTF,*

$$Q^k(\pi, \sigma) = \frac{1}{r_{\pi_1}} \sum_{m=L(\pi^{-1}\sigma)}^{n} P_{\mathbf{r}}[\tilde{X}_{\tilde{C}_{\to m}} = \sigma_{\to m}, \tilde{C}_m < k \leq \tilde{C}_{m+1}, \tilde{X}_k = \pi_1]$$

(3.1)

for $k \geq 0$, with the convention that if $\pi = \sigma$, the summand for $m = 0$ in (3.1) equals r_{π_1} if $k = 0$ and vanishes otherwise.

Proof. Let $Q^k(\pi, \sigma; m)$ denote the probability, starting in a fixed π, that k requests move exactly m distinct records to the front of the list and result in the permutation σ. Thus

$$Q^k(\pi, \sigma) = \sum_{m=0}^{n} Q^k(\pi, \sigma; m).$$

Note that $Q^k(\pi, \sigma; m)$ vanishes if $L(\pi^{-1}\sigma) > m$. If $m \geq 1$ and $L(\pi^{-1}\sigma) \leq m$, then by noting that the first m records must have their last requests occur in the order $\sigma_m, \ldots, \sigma_1$, and conditioning on the times of these requests, we find

$$
(3.2) \quad
\begin{aligned}
Q^k(\pi, \sigma; m) = \sum_{j \to m} &\left[\prod_{v=1}^{m-1} \sum_{i=1}^{v} R(\sigma_{v+1}, \sigma_i) R^{j_v}_{\sigma[v]}(\sigma_i, \sigma_v) \right] \\
&\times \sum_{i=1}^{m} R(\pi_1, \sigma_i) R^{j_m}_{\sigma[m]}(\sigma_i, \sigma_m),
\end{aligned}
$$

where the outer sum is over all m-tuples $j_{\to m} = (j_1, \ldots, j_m)$ whose elements are nonnegative integers summing to $k - m$. The quantity $R^{j_v}_{\sigma[v]}(\sigma_i, \sigma_v)$ is the "taboo" probability of moving from state σ_i to σ_v in j_v steps while hitting only states in $\sigma[v]$.

Let $\rho = r_{\sigma_1}/r_{\pi_1}$. Passing to the time-reversed matrix, (3.2) equals

$$
\begin{aligned}
&\sum_{j \to m} \left[\prod_{v=1}^{m-1} \frac{r_{\sigma_v}}{r_{\sigma_{v+1}}} \sum_{i=1}^{v} \widetilde{R}^{j_v}_{\sigma[v]}(\sigma_v, \sigma_i) \widetilde{R}(\sigma_i, \sigma_{v+1}) \right] \frac{r_{\sigma_m}}{r_{\pi_1}} \sum_{i=1}^{m} \widetilde{R}^{j_m}_{\sigma[m]}(\sigma_m, \sigma_i) \widetilde{R}(\sigma_i, \pi_1) \\
(3.3) =\; &\rho \sum_{j \to m} \left[\prod_{v=1}^{m-1} \sum_{i=1}^{v} \widetilde{R}^{j_v}_{\sigma[v]}(\sigma_v, \sigma_i) \widetilde{R}(\sigma_i, \sigma_{v+1}) \right] \sum_{i=1}^{m} \widetilde{R}^{j_m}_{\sigma[m]}(\sigma_m, \sigma_i) \widetilde{R}(\sigma_i, \pi_1) \\
=\; &\begin{cases} \rho P_{\sigma_1}[\widetilde{X}_{\widetilde{C}_{\to m}} = \sigma_{\to m}, \; \widetilde{C}_m < k, \; \widetilde{C}_{m+1} = k, \; \widetilde{X}_k = \pi_1], & \text{if } \pi_1 \notin \sigma[m] \\ \rho P_{\sigma_1}[\widetilde{X}_{\widetilde{C}_{\to m}} = \sigma_{\to m}, \; \widetilde{C}_m < k, \; \widetilde{C}_{m+1} > k, \; \widetilde{X}_k = \pi_1], & \text{if } \pi_1 \in \sigma[m] \end{cases} \\
(3.4) \quad &= \rho P_{\sigma_1}[\widetilde{X}_{\widetilde{C}_{\to m}} = \sigma_{\to m}, \; \widetilde{C}_m < k \leq \widetilde{C}_{m+1}, \; \widetilde{X}_k = \pi_1].
\end{aligned}
$$

The result follows.   □

Remarks:

1. We can recapture Fill's (1995) formula for the k-step transition probabilities for MTF from (3.3). For fixed $\sigma \in S_n$, let $w_i := r_{\sigma_i}$ and $w_v^+ := \sum_{i=1}^{v} w_i$ with the convention that $w_0^+ := 0$. Note that in the i.i.d. case

$$
\widetilde{R}^{j_v}_{\sigma[v]}(\sigma_v, \sigma_i) = (w_v^+)^{j_v - 1} w_i
$$

for $1 \leq i \leq v$. Hence

$$
\begin{aligned}
&\left[\prod_{v=1}^{m-1} \sum_{i=1}^{v} \widetilde{R}^{j_v}_{\sigma[v]}(\sigma_v, \sigma_i) \widetilde{R}(\sigma_i, \sigma_{v+1}) \right] \sum_{i=1}^{m} \widetilde{R}^{j_m}_{\sigma[m]}(\sigma_m, \sigma_i) \widetilde{R}(\sigma_i, \pi_1) \\
=\; &\left[\prod_{v=1}^{m-1} \sum_{i=1}^{v} (w_v^+)^{j_v - 1} w_i w_{v+1} \right] \sum_{i=1}^{m} (w_m^+)^{j_m - 1} w_i r_{\pi_1} \\
=\; &r_{\pi_1} \left(\prod_{v=2}^{m} w_v \right) \prod_{v=1}^{m} (w_v^+)^{j_v}.
\end{aligned}
$$

Thus from (3.3),

$$Q^k(\pi, \sigma) = \sum_{m=L(\pi^{-1}\sigma)}^{n} \left(\prod_{v=1}^{m} w_v\right) \sum_{j \to m} \prod_{v=1}^{m} (w_v^+)^{j_v}.$$

This calculation is the basis of the proof of Theorem 2.1 in Fill (1995).

2. For moderate n and a specific matrix R, the transition probabilities for MMTF can be computed directly using the formula in Theorem 3.1. The formula's usefulness can be appreciated, for instance, in the case of a 10-element list. The MMTF transition matrix is $3,628,800 \times 3,628,800$. The formula reduces the computations to calculations on matrices of size at most 10×10.

4. Stationary distribution. The main result of this section (Theorem 4.1) gives three representations of the stationary distribution for MMTF. One has a straightforward probabilistic interpretation, while the other two are more suited for numerical calculation in terms of the request matrix or its time reversal.

THEOREM 4.1. *Let Q be the transition matrix for MMTF. Then, for any $\pi, \sigma \in S_n$,*

$$Q^k(\pi, \sigma) \to Q^{\infty}(\sigma) \text{ as } k \to \infty,$$

where

$$(4.1) \quad Q^{\infty}(\sigma) = P_{\mathbf{r}}[\widetilde{X}_{\widetilde{C}_{\to n}} = \sigma]$$

$$(4.2) \qquad\qquad = r_{\sigma_n} \prod_{v=1}^{n-1} \left[\sum_{i=1}^{v} R(\sigma_{v+1}, \sigma_i)(I_{\sigma[v]} - R_{\sigma[v]})^{-1}(\sigma_i, \sigma_v)\right]$$

$$(4.3) \qquad\qquad = r_{\sigma_1} \prod_{v=1}^{n-1} \left[\sum_{i=1}^{v} (I_{\sigma[v]} - \widetilde{R}_{\sigma[v]})^{-1}(\sigma_v, \sigma_i)\widetilde{R}(\sigma_i, \sigma_{v+1})\right],$$

with I the $n \times n$ identity matrix.

Proof. Let $k \to \infty$ in (3.1). All the terms on the right vanish in the limit except for the term with $m = n$, and a simple proof using the strong Markov property and the bounded convergence theorem shows that that term converges to the right side of (4.1).

Note that

$$P_{\mathbf{r}}[\widetilde{X}_{\widetilde{C}_{\to n}} = \sigma] = r_{\sigma_1} P_{\sigma_1}[\widetilde{X}_{\widetilde{C}_{\to (n-1)}} = \sigma_{\to (n-1)}].$$

Expanding this last expression in terms of the $\widetilde{R}$ matrix gives

$$(4.4) \quad Q^{\infty}(\sigma) = r_{\sigma_1} \sum_{j \to (n-1)} \prod_{v=1}^{n-1} \left[\sum_{i=1}^{v} \widetilde{R}_{\sigma[v]}^{j_v}(\sigma_v, \sigma_i)\widetilde{R}(\sigma_i, \sigma_{v+1})\right]$$

$$= r_{\sigma_1} \prod_{v=1}^{n-1} \left[\sum_{i=1}^{v} (I_{\sigma[v]} - \tilde{R}_{\sigma[v]})^{-1}(\sigma_v, \sigma_i)\tilde{R}(\sigma_i, \sigma_{v+1}) \right].$$

The outer sum in (4.4) is over all $(n-1)$-tuples $j_{\to(n-1)} = (j_1, \ldots, j_{n-1})$ whose elements are nonnegative integers. This proves (4.3), and (4.2) follows immediately from the relationship between R and $\tilde{R}$. $\quad\Box$

Remark: In the i.i.d. case, it follows from (4.1) that for $\sigma \in S_n$, $Q^\infty(\sigma)$ is the probability of obtaining the ordered list $(\sigma_1, \ldots, \sigma_n)$ by sampling without replacement from $\{\sigma_1, \ldots, \sigma_n\}$. Thus, in the notation $w_i = r_{\sigma_i}$ and $w_v^+ = \sum_{i=1}^{v} w_i$ introduced earlier,

$$Q^\infty(\sigma) = \frac{w_1 \cdots w_n}{\prod_{j=0}^{n-1}(1 - w_j^+)}.$$

As pointed out in Section 2, the ergodicity of the R-chain is insufficient for the Q-chain to be ergodic. However, there always exists a unique positive recurrent communication class RC, with $Q^\infty(\sigma) > 0$ if $\sigma \in RC$ and $Q^\infty(\sigma) = 0$ if σ is transient, that is, if $\sigma \notin RC$. The following lemma gives a necessary and sufficient condition for $\sigma \in RC$.

LEMMA 4.1. *Let $\sigma \in S_n$. Then $Q^\infty(\sigma) > 0$ if and only if for each $v \in [n-1]$ there exists $k \geq 1$ with*

$$(4.5) \qquad \tilde{R}_{\sigma[v+1]}^k(\sigma_v, \sigma_{v+1}) > 0.$$

Proof. Clearly, (4.5) holds for some $k \geq 1$ if and only if

$$\sum_{i=1}^{v} \tilde{R}_{\sigma[v]}^j(\sigma_v, \sigma_i)\tilde{R}(\sigma_i, \sigma_{v+1}) > 0$$

for some $j \geq 0$. The lemma then follows from (4.4). $\quad\Box$

In the remainder of this section we consider the birth-and-death case introduced in Section 2. Let R be an ergodic chain with transitions

$$(4.6) \qquad R(i,j) = \begin{cases} q_i, & \text{if } j = i-1 \\ 1 - q_i - p_i, & \text{if } j = i \\ p_i, & \text{if } j = i+1 \\ 0, & \text{otherwise} \end{cases}$$

for $1 \leq i, j \leq n$, with $q_1 = p_n = 0$. By using results for hitting times for birth-and-death chains one can compute $Q^\infty(\sigma)$ explicitly.

THEOREM 4.2. *Let R be as defined in (4.6). Given $\sigma \in S_n$, a necessary and sufficient condition for $Q^\infty(\sigma) > 0$ is that $(\sigma_2, \ldots, \sigma_n)$ be an interleaving of the two sequences $(\sigma_1 - 1, \sigma_1 - 2, \ldots, 1)$ and $(\sigma_1 + 1, \sigma_1 + 2, \ldots, n)$. For such a σ,*

$$(4.7) \qquad Q^\infty(\sigma) = \frac{\hat{r}_{\sigma_1}}{\sum_{i=1}^{n} \hat{r}_i} \prod_{j=1}^{n-2} h(\sigma; j),$$

where $h(\sigma; j)$ is defined to be

$$
\begin{cases}
1, & \text{if } \min\sigma[j] = 1 \text{ or } \max\sigma[j] = n \\
\dfrac{\gamma(\min\sigma[j]-1,\sigma_j-1)}{\gamma(\min\sigma[j]-1,\sigma_{j+1}-1)}, & \text{if } \sigma_j < \sigma_{j+1},\ 1 < \min\sigma[j],\ \text{and } \max\sigma[j] < n \\
\dfrac{\gamma(\sigma_j,\max\sigma[j])}{\gamma(\sigma_{j+1},\max\sigma[j])}, & \text{if } \sigma_j > \sigma_{j+1},\ 1 < \min\sigma[j],\ \text{and } \max\sigma[j] < n,
\end{cases}
$$

(4.8)
$$
\gamma(a, b) := \sum_{i=a}^{b} \prod_{j=1}^{i} \frac{q_j}{p_j} \quad \text{for } 1 \le a \le b < n,
$$

and

(4.9)
$$
\hat{r}_j := \begin{cases} 1, & j = 1 \\ \dfrac{p_1 \cdots p_{j-1}}{q_2 \cdots q_j}, & 1 < j \le n. \end{cases}
$$

Proof. Let $\tilde{T}_i$ be the first hitting time of state i for $\tilde{X}$. That is,

(4.10)
$$
\tilde{T}_i := \inf\{k \ge 0 : \tilde{X}_k = i\}.
$$

Similarly define $\tilde{T}_A$ for a subset A of the state space. For $1 \le j \le n - 2$, let

(4.11)
$$
\hat{h}(\sigma; j) := P_{\sigma_j}[\tilde{T}_{\sigma_{j+1}} < \tilde{T}_{\{\sigma_{j+2},\ldots,\sigma_n\}}].
$$

Then

$$
Q^\infty(\sigma) = r_{\sigma_1} \prod_{j=1}^{n-2} \hat{h}(\sigma; j).
$$

A consequence of Lemma 4.1 is that $Q^\infty(\sigma) > 0$ if and only if $(\sigma_2, \ldots, \sigma_n)$ is an interleaving of the two sequences

$$
(\sigma_1 - 1, \sigma_1 - 2, \ldots, 1) \text{ and } (\sigma_1 + 1, \sigma_1 + 2, \ldots, n).
$$

For such σ,

$$
\hat{h}(\sigma; j) = \begin{cases}
P_{\sigma_j}[\tilde{T}_{\sigma_{j+1}} < \tilde{T}_{(\min\sigma[j])-1}], & \text{if } \sigma_j < \sigma_{j+1} \\
P_{\sigma_j}[\tilde{T}_{\sigma_{j+1}} < \tilde{T}_{(\max\sigma[j])+1}], & \text{if } \sigma_j > \sigma_{j+1}
\end{cases}
$$

with the natural convention that $\hat{h}(\sigma; j) = 1$ if $\min\sigma[j] = 1$ or $\max\sigma[j] = n$. It is elementary and well known that for birth-and-death chains

$$
P_x[\tilde{T}_b < \tilde{T}_a] = \frac{\sum_{y=a}^{x-1} \gamma_y}{\sum_{y=a}^{b-1} \gamma_y}, \quad a < x < b,
$$

where

$$\gamma_y := \frac{q_1 \cdots q_y}{p_1 \cdots p_y}.$$

Thus $\hat{h}(\sigma; j)$ equals $h(\sigma; j)$ in the statement of the theorem. The stationary distribution for birth-and-death chains is well known, giving

$$r_{\sigma_1} = \frac{\hat{r}_{\sigma_1}}{\sum_{i=1}^{n} \hat{r}_i},$$

with $\hat{r}$ defined at (4.9).   □

We now specialize to a simple case for which we can obtain completely explicit results. In the following corollary we treat the case of a request chain whose stationary distribution $\mathbf{r}$ is uniform on $[n]$. This is the same stationary distribution as the request chain for MTF with equal weights. But, as we'll see, the MMTF chain exhibits behavior quite different from that of the i.i.d. case.

COROLLARY 4.1. *Let the R-chain be the following simple symmetric random walk on* $[n]$: *For fixed* $0 < p \leq 1/2$, $p_i = q_{i+1} = p$ *for* $i = 1, \ldots, n-1$. *Then for positive recurrent states* $\sigma \in S_n$,

$$Q^{\infty}(\sigma) = \frac{1}{n\alpha!} \prod_{1 \leq j \leq \alpha:\ \sigma_j \sim \sigma_{j+1}} j,$$

where $\alpha = \min\{\sigma^{-1}(1), \sigma^{-1}(n)\}$ *and* $a \sim b$ *means* a *is adjacent to* b, *that is,* $b \in \{a-1, a+1\}$.

Proof. For simple symmetric random walk,

(4.12) $\gamma(a, b) = b - a + 1, \quad 1 \leq a < b \leq n.$

Thus $h(\sigma; j)$ equals

$$(4.13) \quad \begin{cases} 1, & \text{if } \min \sigma[j] = 1 \text{ or } \max \sigma[j] = n \\ \frac{\sigma_j - \min \sigma[j] + 1}{\sigma_{j+1} - \min \sigma[j] + 1}, & \text{if } \sigma_j < \sigma_{j+1},\ 1 < \min \sigma[j], \text{ and } \max \sigma[j] < n \\ \frac{\max \sigma[j] - \sigma_j + 1}{\max \sigma[j] - \sigma_{j+1} + 1}, & \text{if } \sigma_j > \sigma_{j+1},\ 1 < \min \sigma[j], \text{ and } \max \sigma[j] < n. \end{cases}$$

Further, since σ is an interleaving,

$$(4.14) \quad \text{if } \left\{ \begin{array}{c} \sigma_j < \sigma_{j+1} \\ \sigma_j > \sigma_{j+1} \end{array} \right\} \text{ then } \left\{ \begin{array}{c} \sigma_{j+1} = \max \sigma[j] + 1 \\ \sigma_{j+1} = \min \sigma[j] - 1 \end{array} \right\}.$$

Also, if $\sigma_j < \sigma_{j+1}$ then

$$(4.15) \quad \left\{ \begin{array}{c} \sigma_j = \max \sigma[j] \\ \sigma_j = \min \sigma[j] \end{array} \right\} \text{ if } \left\{ \begin{array}{c} \sigma_j \sim \sigma_{j+1} \\ \sigma_j \nsim \sigma_{j+1} \end{array} \right\}$$

and similarly if $\sigma_j > \sigma_{j+1}$ with the roles of $\max \sigma[j]$ and $\min \sigma[j]$ reversed. From (4.13), (4.14), and (4.15) it follows that

$$(4.16) \qquad h(\sigma; j) = \begin{cases} j/(j+1), & \text{if } \sigma_j \sim \sigma_{j+1} \\ 1/(j+1), & \text{if } \sigma_j \not\sim \sigma_{j+1}. \end{cases}$$

The result is a consequence of (4.16) and the fact that $\hat{r}_j = 1$ for $1 \le j \le n$. □

Remarks:

1. In the setting of Corollary 4.1, $Q^\infty(\sigma)$ is maximized by $\sigma = \text{id}$ and $\sigma = \text{rev}$ (and only these), with $Q^\infty(\max) = 1/n$; $Q^\infty(\sigma)$, when positive, is minimized by exactly two permutations σ (e.g., by

$$\sigma = \left(\frac{n}{2}, \frac{n}{2}+1, \frac{n}{2}-1, \frac{n}{2}+2, \frac{n}{2}-2, \dots, 1, n \right),$$

if n is even), with $Q^\infty(\min) = 1/n!$. There are $2^{n-1} \ll n!$ permutations $\sigma \in S_n$ with $Q^\infty(\sigma) > 0$.

2. With the simple symmetric random walk request chain discussed above, MMTF behaves quite differently from MTF with equal weights, even though both request chains have the same stationary distribution. This is underscored by the magnitude of stationary expected search cost ESC. In the random walk case, using the main result of Lam et al. (1984) one finds $\text{ESC} = 1 + 2p(H_n - 1)$, where $H_n = \sum_{k=1}^{n} k^{-1}$. In the MTF case, $\text{ESC} = (n+1)/2$.

5. Spectral analysis. Phatarfod and Dyte (1993) determined the eigenvalues, with their multiplicities, for MMTF. We show how to derive these from Theorem 1 and go a step further by explicitly giving the associated idempotents. For $S \subseteq [n]$, we write $\lambda_1(S), \lambda_2(S), \dots, \lambda_{|S|}(S)$ for the eigenvalues of the principal submatrix R_S of the request matrix R. We will use the notation $1(A)$ for the indicator of A.

THEOREM 5.1. *(a) The set of eigenvalues for the MMTF transition matrix Q defined at (2.2) is the set of all eigenvalues of all the principal submatrices R_S with $|S| \ne 0, n-1$ of the request chain R. For $S \subseteq [n]$, the eigenvalue $\lambda_i(S)$ corresponding to the ith eigenvalue of R_S has multiplicity in Q equal to $D_{n-|S|}$, the number of derangements (permutations with no fixed points) of $n - |S|$ objects.*

(b) If all the eigenvalues of all of the principal submatrices of R are distinct and nonzero, then MMTF is diagonalizable, with spectral decomposition

$$(5.1) \qquad Q = \sum_{\substack{S \subseteq [n] \\ |S| \ne 0, n-1}} \sum_{i=1}^{|S|} \lambda_i(S) E_{i,S}.$$

Here $E_{i,S}$ is the principal idempotent

$$(5.2) \quad E_{i,S}(\pi,\sigma) := 1(\sigma[|S|] = S) \sum_{\substack{m=|S| \vee L(\pi^{-1}\sigma)}}^{n} \sum_{\substack{t_{\to m} \\ t_{|S|}=i}} \frac{H(t_{\to m}, \sigma_{\to m}, \pi_1)}{b_{|S|}(t_{\to m}, \sigma_{\to m})},$$

where the inner sum in (5.2) is over all m-tuples $t_{\to m} = (t_1, \ldots, t_m)$ of integers such that $1 \le t_k \le k$ for $1 \le k \le m$ and $t_{|S|} = i$. Also

$$b_x(t_{\to j}, \sigma_{\to j}) := \prod_{\substack{i \ne x \\ 0 \le i \le j}} (\lambda_{t_x}(\sigma[x]) - \lambda_{t_i}(\sigma[i]))$$

with $t_0 := 0$ and $\lambda_0(\emptyset) := 0$, and

$$H(t_{\to j}, \sigma_{\to j}, x) := \left[\prod_{v=1}^{j-1} \sum_{i=1}^{v} R(\sigma_{v+1}, \sigma_i) F_{t_v, \sigma[v]}(\sigma_i, \sigma_v) \right] \sum_{i=1}^{j} R(x, \sigma_i) F_{t_j, \sigma[j]}(\sigma_i, \sigma_j),$$

where $F_{i,S}$ is the principal idempotent of R_S corresponding to $\lambda_i(S)$.

Proof. (a) By Theorem 3.1, for $k \ge 1$,

$$(5.3) \quad Q^k(\pi, \pi) = \frac{1}{r_{\pi_1}} \sum_{m=1}^{n} \Pr[\tilde{X}_{\tilde{C}_{\to m}} = \pi_{\to m}, \; \tilde{C}_m \le k < \tilde{C}_{m+1}, \; \tilde{X}_k = \pi_1].$$

Thus

$$\operatorname{tr}(Q^k) = \sum_{\pi \in S_n} Q^k(\pi, \pi)$$

$$= \sum_{m=1}^{n} (n-m)! \sum_{\pi_{\to m}} \frac{1}{r_{\pi_1}} \Pr[\tilde{X}_{\tilde{C}_{\to m}} = \pi_{\to m}, \; \tilde{C}_m \le k < \tilde{C}_{m+1}, \; \tilde{X}_k = \pi_1]$$

$$= \sum_{m=1}^{n} (n-m)! \sum_{S \in \binom{[n]}{m}} \sum_{x \in S} P_x[[\tilde{X}_{\tilde{C}_{\to m}}] = S, \; \tilde{C}_m \le k < \tilde{C}_{m+1}, \; \tilde{X}_k = x]$$

$$= \sum_{m=1}^{n} (n-m)! \sum_{S \in \binom{[n]}{m}} \sum_{x \in S} P_x[[\tilde{X}_{\to k}] = S, \; \tilde{X}_k = x],$$

where the sum $\sum_{S \in \binom{[n]}{m}}$ is over all m-element subsets S of $[n]$. By inclusion-exclusion, for $x \in S$ we have

$$P_x[[\tilde{X}_{\to k}] = S, \; \tilde{X}_k = x] = \sum_{T \subseteq S} (-1)^{|S|-|T|} P_x[[\tilde{X}_{\to k}] \subseteq T, \; \tilde{X}_k = x]$$

$$= \sum_{T \subseteq S} (-1)^{|S|-|T|} 1(x \in T) \tilde{R}_T^k(x, x).$$

Therefore,

$$
\begin{aligned}
\operatorname{tr}(Q^k) &= \sum_{m=1}^{n}(n-m)! \sum_{S\in\binom{[n]}{m}} \sum_{T\subseteq S}(-1)^{|S|-|T|} \sum_{x\in T} \widetilde{R}_T^k(x,x) \\
&= \sum_{m=1}^{n}(n-m)! \sum_{S\in\binom{[n]}{m}} \sum_{T\subseteq S}(-1)^{|S|-|T|}\operatorname{tr}(\widetilde{R}_T^k) \\
&= \sum_{\emptyset\neq T\subseteq[n]} \operatorname{tr}(\widetilde{R}_T^k) \sum_{S\supseteq T}(-1)^{|S|-|T|}(n-|S|)! \\
(5.4) \qquad &= \sum_{\emptyset\neq T\subseteq[n]} D_{n-|T|}\operatorname{tr}(\widetilde{R}_T^k).
\end{aligned}
$$

It is not hard to check that (5.4) also holds for $k = 0$. Since $\operatorname{tr}(R_T^k) = \operatorname{tr}(\widetilde{R}_T^k)$, part (a) of Theorem 5.1 follows.

For part (b) assume that the principal submatrices of R have spectral decompositions given by

$$
R_S = \sum_{i=1}^{|S|} \lambda_i(S) F_{i,S}, \quad \text{for } \emptyset \neq S \subseteq [n].
$$

Until (5.6) we use for abbreviation the convention that $\sigma_{m+1} := \pi_1$. From (3.2) we have

$$
\begin{aligned}
Q^k(\pi,\sigma;m) &= \sum_{j\to m}\left[\prod_{v=1}^{m}\sum_{i=1}^{v} R(\sigma_{v+1},\sigma_i)R_{\sigma[v]}^{j_v}(\sigma_i,\sigma_v)\right] \\
&= \sum_{j\to m}\left[\prod_{v=1}^{m}\sum_{i=1}^{v} R(\sigma_{v+1},\sigma_i)\sum_{t=1}^{v}\lambda_t^{j_v}(\sigma[v])F_{t,\sigma[v]}(\sigma_i,\sigma_v)\right] \\
(5.5)\qquad &= \sum_{j\to m}\sum_{t\to m}\left[\prod_{v=1}^{m}\sum_{i=1}^{v} R(\sigma_{v+1},\sigma_i)\lambda_{t_v}^{j_v}(\sigma[v])F_{t_v,\sigma[v]}(\sigma_i,\sigma_v)\right],
\end{aligned}
$$

where the second sum is over integer m-tuples $t_{\to m} = (t_1,\ldots,t_m)$ with $1 \le t_v \le v$ for $1 \le v \le m$. Continuing, (5.5) equals

$$
\begin{aligned}
&\sum_{t\to m}\left[\sum_{j\to m}\prod_{v=1}^{m}\lambda_{t_v}^{j_v}(\sigma[v])\right]\prod_{v=1}^{m}\sum_{i=1}^{v}R(\sigma_{v+1},\sigma_i)F_{t_v,\sigma[v]}(\sigma_i,\sigma_v) \\
(5.6)\qquad &= \sum_{t\to m}\sum_{z=0}^{m}\frac{\lambda_{t_z}^k(\sigma[z])}{b_z(t_{\to m},\sigma_{\to m})}H(t_{\to m},\sigma_{\to m},\pi_1),
\end{aligned}
$$

where for the last equality we have used Proposition A.1 in Fill (1995), with $t_0 := 0$, $\lambda_0(\emptyset) := 0$, and $b_z(t_{\to m},\sigma_{\to m})$ and $H(t_{\to m},\sigma_{\to m},\pi_1)$ as defined in the statement of the theorem. Thus

$$(5.7) \quad Q^k(\pi, \sigma) = \sum_{m=L(\pi^{-1}\sigma)}^{n} \sum_{t_{\to m}} \sum_{z=0}^{m} \frac{\lambda_{t_z}^k(\sigma[z])}{b_z(t_{\to m}, \sigma_{\to m})} H(t_{\to m}, \sigma_{\to m}, \pi_1),$$

and by rearrangement and the fact (from (a)) that neither 0 nor any $\lambda_i(S)$ with $|S| = n - 1$ is an eigenvalue of Q,

$$Q^k(\pi, \sigma)$$

$$(5.8) \quad = \sum_{|\dot{s}|^{-n}} \sum_{z=0}^{n} \lambda_{t_z}^k(\sigma[z]) \sum_{m=z \vee L(\pi^{-1}\sigma_n)}^{n} \frac{m! H(t_{\to m}, \sigma_{\to m}, \pi_1)}{n! b_z(t_{\to m}, \sigma_{\to m})}$$

$$= \sum_{\substack{S \subseteq [n] \\ |S| \neq 0, n-1}} \sum_{i=1}^{n} \lambda_i^k(S) \mathbf{1}(\sigma[|S|] = S) \sum_{m=|S| \vee L(\pi^{-1}\sigma)} \sum_{\substack{t_{\to m} \\ t_{|S|} = i}} \frac{H(t_{\to m}, \sigma_{\to m}, \pi_1)}{b_{|S|}(t_{\to m}, \sigma_{\to m})},$$

where the outer sum in (5.8) is over all n-tuples $t_{\to n} = (t_1, \ldots, t_n)$ of integers with $1 \leq t_z \leq z$ for $1 \leq z \leq n$. $\square$

Remarks:

1. We stress that Theorem 4(a) makes *no* assumption on the diagonalizability or the eigenvalues of the request chain. The assumption on the eigenvalues in part (b) is for simplicity and convenience. More general cases can be handled with a slight modification of the proof using a perturbation argument.

2. In the i.i.d. case, Theorem 5.1 recovers results in Fill (1995). In particular, the set of eigenvalues for MTF are all numbers of the form

$$\lambda_S := \sum_{i \in S} r_i$$

with $S \subseteq [n]$ and $|S| \neq n - 1$.

6. Separation

6.1. General result. In Sections 6 and 7 we consider two common notions of discrepancy between the distribution of a Markov chain at a fixed time and its stationary distribution. In this subsection we derive an upper bound on separation for MMTF in terms of the separation for the request chain, in the next subsection we apply our result to two examples, and in Section 7 we treat total variation distance. For background on separation see Aldous and Diaconis (1986, 1987). These authors assume ergodicity, which implies that the stationary distribution is unique and strictly positive. We extend the usual definition somewhat.

For a finite-state Markov chain $(Y_n)_{n \geq 0}$ with transition matrix P and unique stationary distribution P^∞, let

$$s_{i,j}(k; P) := \begin{cases} 1 - \frac{P^k(i,j)}{P^\infty(j)}, & \text{if } P^\infty(j) > 0 \\ 0, & \text{otherwise} \end{cases}$$

and define

$$\text{sep}_i(k; P) := \max_j s_{i,j}(k; P)$$

to be the *separation* of the Y-chain at time k when started in state i. We say that a state j^* *achieves* the separation $\text{sep}_i(k; P)$ if $P^\infty(j^*) > 0$ and

$$\text{sep}_i(k; P) = 1 - \frac{P^k(i, j^*)}{P^\infty(j^*)}.$$

We write

$$\text{sep}^*(k; P) := \max_i \text{sep}_i(k; P) = \max_{i,j} s_{i,j}(k; P)$$

for the worst-case separation.

The notion of strong stationary time gives a probabilistic approach to bounding speed of convergence to stationarity for Markov chains. A *strong stationary time* for a Markov chain Y with transition matrix P and unique stationary distribution P^∞, started in state i, is a randomized stopping time T such that Y_T has the distribution P^∞ and is independent of T. Separation can be bounded above by the tail probabilities of a strong stationary time. More precisely,

$$(6.1) \qquad \text{sep}_i(k; P) \le P[T > k]$$

for $k \ge 0$. A fastest strong stationary time, or *time to stationarity*, is a strong stationary time which achieves equality in (6.1) for all $k \ge 0$. See Diaconis and Fill (1990) for further discussion.

We next review standard terminology in the study of convergence to stationarity. We say that $k = k(n, c)$ steps are sufficient for convergence to stationarity in separation if there exists a function H, independent of n, such that $\text{sep}^*(k; P) \le H(c)$ and $H(c) \to 0$ as $c \to \infty$. We say that $k = k(n, c)$ steps are necessary for convergence to stationarity in separation if there exists a function h, independent of n, such that $\text{sep}^*(k; P) \ge h(c)$ and $h(c) \to 1$ as c tends to $-\infty$ (or whatever the infimum of possible values of c might be.) If $g(n) = o(f(n))$ and $k(n, c) = f(n) + cg(n)$ steps are necessary and sufficient, we say that a "cutoff" occurs at time $f(n)$. Analogous definitions can be given for convergence to stationarity in total variation distance.

The following theorem provides an upper bound on separation for MMTF (Q) in terms of the separation for the request chain (R).

THEOREM 6.1. *Let* $\pi \in S_n$. *For* $\sigma \in S_n$ *with* $Q^\infty(\sigma) > 0$, *let* $T(\sigma)$ *be a random variable whose distribution is the same as the conditional distribution of* $\widetilde{C}_{n-1}$ *given* $\widetilde{X}_{\widetilde{C}_{\to n}} = \sigma$. *Further, for* $j \in [n]$ *let* S_j *be a random variable, independent of* $T(\sigma)$, *whose distribution is that of a fastest strong stationary time for the request chain started in state* j. *Then*

$$(6.2) \qquad s_{\pi,\sigma}(k; Q) \le P[S_{\pi_1} + T(\sigma) > k], \; k \ge 0.$$

Equality is achieved in (6.2) for all $k \geq 0$ if and only if $L(\pi^{-1}\sigma) = n - 1$ and state σ_{n-1} achieves the separation $\mathrm{sep}_{\pi_1}(k; R)$ for every $k > 0$. If these conditions hold and also σ maximizes $\mathcal{L}(T(\sigma))$ stochastically, then

$$\mathrm{sep}_\pi(k; Q) = P[S_{\pi_1} + T(\sigma) > k], \quad k \geq 0.$$

Proof. Let $\pi, \sigma \in S_n$ and $\rho = r_{\sigma_1}/r_{\pi_1}$. From Theorem 3.1,

$$
\begin{aligned}
Q^k(\pi, \sigma) \;\geq\;& \rho \sum_{m=n-1}^{n} P_{\sigma_1}[\tilde{X}_{\tilde{C}_{\rightarrow m}} = \sigma_{\rightarrow m}, \tilde{C}_m < k \leq \tilde{C}_{m+1}, \tilde{X}_k = \pi_1] \\
=\;& \rho\{P_{\sigma_1}[\tilde{X}_{\tilde{C}_{\rightarrow(n-1)}} = \sigma_{\rightarrow(n-1)}, \tilde{C}_{n-1} < k \leq \tilde{C}_n, \tilde{X}_k = \pi_1] \\
& + P_{\sigma_1}[\tilde{X}_{\tilde{C}_{\rightarrow n}} = \sigma, \tilde{C}_n < k, \tilde{X}_k = \pi_1]\} \\
=\;& \rho P_{\sigma_1}[\tilde{X}_{\tilde{C}_{\rightarrow(n-1)}} = \sigma_{\rightarrow(n-1)}, \tilde{C}_{n-1} \leq k - 1, \tilde{X}_k = \pi_1] \\
=\;& \rho \sum_{t=0}^{k-1} P_{\sigma_1}[\tilde{X}_{\tilde{C}_{\rightarrow(n-1)}} = \sigma_{\rightarrow(n-1)}, \tilde{C}_{n-1} = t]\tilde{R}^{k-t}(\sigma_{n-1}, \pi_1) \\
=\;& \rho \sum_{t=0}^{k-1} P_{\sigma_1}[\tilde{X}_{\tilde{C}_{\rightarrow(n-1)}} = \sigma_{\rightarrow(n-1)}, \tilde{C}_{n-1} = t]\frac{r_{\pi_1}}{r_{\sigma_{n-1}}}R^{k-t}(\pi_1, \sigma_{n-1}) \\
\geq\;& r_{\sigma_1} \sum_{t=0}^{k-1} P_{\sigma_1}[\tilde{X}_{\tilde{C}_{\rightarrow(n-1)}} = \sigma_{\rightarrow(n-1)}, \tilde{C}_{n-1} = t](1 - \mathrm{sep}_{\pi_1}(k - t; R)) \\
=\;& r_{\sigma_1} P_{\sigma_1}[\tilde{X}_{\tilde{C}_{\rightarrow n}} = \sigma] \sum_{t=0}^{k-1} P[T(\sigma) = t](1 - \mathrm{sep}_{\pi_1}(k - t; R)) \\
=\;& Q^\infty(\sigma) \sum_{t=0}^{k} P[T(\sigma) = t](1 - \mathrm{sep}_{\pi_1}(k - t; R)) \\
=\;& Q^\infty(\sigma) \sum_{t=0}^{k} P[T(\sigma) = t]P[S_{\pi_1} \leq k - t],
\end{aligned}
$$

where $T(\sigma)$ and S_{π_1} are as defined in the statement of the theorem. The penultimate equality holds since $\mathrm{sep}_{\pi_1}(0; R) = 1$, assuming $n \geq 2$. For $k \geq n - 1$ the first inequality becomes an equality if and only if $L(\pi^{-1}\sigma) = n - 1$ and the second inequality is an equality if and only if state σ_{n-1} achieves the separation $\mathrm{sep}_{\pi_1}(u; R)$ at every strictly positive time u. Since S_{π_1} and $T(\sigma)$ are taken to be independent, the result follows. $\square$

6.2. Examples. *Example 1. I.i.d. case.* In the i.i.d. case $S_{\pi_1} \equiv 1$ for all π. Recalling the notation in the first remark of Section 3, the probability generating function for $T(\sigma)$ is given by

$$\sum_{t=0}^{\infty} P[T(\sigma) = t]z^t = \frac{\sum_{t=0}^{\infty} z^t \sum (w_1^+)^{j_1} w_2(w_2^+)^{j_2} w_3 \cdots (w_{n-2}^+)^{j_{n-2}} w_{n-1}}{\sum (w_1^+)^{j_1} w_2(w_2^+)^{j_2} w_3 \cdots (w_{n-2}^+)^{j_{n-2}} w_{n-1}},$$

where the unmarked sum in the numerator is over all nonnegative $(n-2)$-tuples $j_{\to(n-2)}$ summing to $t-(n-2)$ and the sum in the denominator is over all nonnegative $(n-2)$-tuples $j_{\to(n-2)}$. Interchanging sums, the numerator equals the unrestricted sum

$$z^{n-2}\sum (w_1^+ z)^{j_1} w_2 \cdots (w_{n-2}^+ z)^{j_{n-2}} w_{n-1} = z^{n-2}\frac{w_2 \cdots w_{n-1}}{(1-w_1^+ z)\cdots(1-w_{n-2}^+ z)},$$

while the denominator equals

$$\frac{w_2 \cdots w_{n-1}}{(1-w_1^+)\cdots(1-w_{n-2}^+)}.$$

Thus

$$\sum_{t=0}^{\infty} P[T(\sigma)=t]z^t = \prod_{v=1}^{n-2}\left[\frac{(1-w_v^+)z}{1-w_v^+ z}\right].$$

That is,

(6.3) $$T(\sigma) \sim \oplus_{v=1}^{n-2}\mathrm{Geom}(1-w_v^+),$$

where the notation indicates the convolution of $n-2$ geometric distributions with the indicated parameters. Hence by Theorem 6.1,

(6.4) $$1-\frac{Q^k(\pi,\sigma)}{Q^{\infty}(\sigma)} \le P[\oplus_{v=0}^{n-2}\mathrm{Geom}(1-w_v^+) > k],$$

which extends Lemma 4.6 in Fill (1995).

  In the i.i.d. case the request chain has separation equal to 0 at all positive times. Thus there is equality in (6.4) if and only if $L(\pi^{-1}\sigma) = n-1$, from which follows Theorem 4.1 in Fill (1995), which we reproduce for completeness:

  THEOREM 6.2. *Consider the move-to-front scheme with weights $r_1,\ldots,$ r_n, and suppose (without loss of generality) that $r_1 \ge r_2 \ge \cdots \ge r_n > 0$. Let $\pi \in S_n$ be any permutation with $\pi^{-1}(n-1) > \pi^{-1}(n)$. Then*

$$\mathrm{sep}(k;Q) = 1-\frac{Q^k(\pi,\mathrm{id})}{Q^{\infty}(\mathrm{id})} = P[T^* > k], \quad k=0,1,2,\ldots,$$

where the law of T^* is the convolution of Geometric$(1-r_v^+)$ distributions, $v=0,1,\ldots,n-2$.

  Remark: For general ergodic R we partially generalize the result in (6.3) by exhibiting the distribution of $T(\sigma)(= \mathcal{L}(\widetilde{C}_{n-1}|X_{\widetilde{C}_{\to n}} = \sigma))$, where $Q^{\infty}(\sigma) > 0$, as the convolution of $n-2$ distributions. Write $\widetilde{C}_0 := 0$ and let $\widetilde{W}_v := \tilde{C}_v - \tilde{C}_{v-1}$ for $1 \leq v \leq n$. Thus $\widetilde{W}_v$ is the waiting time from the $(v-1)$st state covered by $\tilde{X}$ to the vth, and

$$\tilde{C}_{n-1} = \sum_{v=1}^{n-1} \widetilde{W}_v = \sum_{v=2}^{n-1} \widetilde{W}_v.$$

But

$$P[\widetilde{W}_2 = w_2, \ldots, \widetilde{W}_{n-1} = w_{n-1} | \tilde{X}_{\tilde{C}_{\to n}} = \sigma]$$

$$= \frac{P_{\mathbf{r}}[\widetilde{W}_2 = w_2, \ldots, \widetilde{W}_{n-1} = w_{n-1}, \tilde{X}_{\tilde{C}_{\to n}} = \sigma]}{P_{\mathbf{r}}[\tilde{X}_{\tilde{C}_{\to n}} = \sigma]}$$

$$= \frac{\prod_{v=1}^{n-2} \sum_{i=1}^{v} \tilde{R}_{\sigma[v]}^{w_{v+1}-1}(\sigma_v, \sigma_i) \tilde{R}(\sigma_i, \sigma_{v+1})}{\prod_{v=1}^{n-2} \sum_{i=1}^{v} (I_{\sigma[v]} - \tilde{R}_{\sigma[v]})^{-1}(\sigma_v, \sigma_i) \tilde{R}(\sigma_i, \sigma_{v+1})}$$

(6.5)
$$= \prod_{v=1}^{n-2} \frac{\sum_{i=1}^{v} \tilde{R}_{\sigma[v]}^{w_{v+1}-1}(\sigma_v, \sigma_i) \tilde{R}(\sigma_i, \sigma_{v+1})}{\sum_{i=1}^{v} (I_{\sigma[v]} - \tilde{R}_{\sigma[v]})^{-1}(\sigma_v, \sigma_i) \tilde{R}(\sigma_i, \sigma_{v+1})}.$$

It follows that $\widetilde{W}_1, \ldots, \widetilde{W}_{n-1}$ are conditionally independent given $\tilde{X}_{\tilde{C}_{\to n}} = \sigma$, with $\widetilde{W}_1 = 0$ and

(6.6) $P[\widetilde{W}_{v+1} = w | \tilde{X}_{\tilde{C}_{\to n}} = \sigma] = \dfrac{\sum_{i=1}^{v} \tilde{R}_{\sigma[v]}^{w-1}(\sigma_v, \sigma_i) \tilde{R}(\sigma_i, \sigma_{v+1})}{\sum_{i=1}^{v} (I_{\sigma[v]} - \tilde{R}_{\sigma[v]})^{-1}(\sigma_v, \sigma_i) \tilde{R}(\sigma_i, \sigma_{v+1})}$

$$= P[\widetilde{W}_{v+1} = w | \tilde{X}_{\tilde{C}_{\to(v+1)}} = \sigma_{\to(v+1)}]$$

for $1 \leq v \leq n-2$.

We remark that in the i.i.d. case, it follows after some calculation from (6.6) that

$$\mathcal{L}(\widetilde{W}_{v+1} | \tilde{X}_{\tilde{C}_{\to n}} = \sigma) = \mathrm{Geom}(1 - w_v^+),$$

as at (6.3).

Example 2. Random walk. In treating the simple symmetric random walk example of Corollary 4.1, considerable technical difficulties arise involving the endpoints 1 and n of the request chain's state space. We therefore suppose instead that the request chain is simple symmetric *circular* random walk on $[n]$. That is, for fixed $0 < p \leq 1/2$,

$$R(i,j) := \begin{cases} p, & \text{if } j \equiv i-1 \pmod{n} \\ 1-2p, & \text{if } j \equiv i \pmod{n} \\ p, & \text{if } j \equiv i+1 \pmod{n} \\ 0, & \text{otherwise.} \end{cases}$$

An analysis like that in the birth-and-death case shows that $Q^\infty(\sigma) > 0$ if and only if $\sigma_{j+1} \sim \sigma[j]$ for $j \in [n-2]$, in which case

$$Q^\infty(\sigma) = \frac{1}{n!} \prod_{1 \le j \le n-2: \, \sigma_j \sim \sigma_{j+1}} j,$$

where now $a \sim b$ means that a is circularly adjacent to b and $a \sim B$ means that $a \sim b$ for some $b \in B$.

THEOREM 6.3. *Consider MMTF whose request chain R is the simple symmetric circular random walk described above. For simplicity assume that $n = 2m$ is even and $0 < p \le 1/4$.*

(a) A cutoff for separation occurs at time $(18p)^{-1}n^3$: $(18p)^{-1}n^3 + cn^{5/2}$ steps are necessary and sufficient for convergence to stationarity in separation.

(b) Let $\sigma^* = (2m, 1, 2m-1, 2, \ldots, m+2, m-1, m+1, m)$. Then the separation $\mathrm{sep}_{\mathrm{id}}(k; Q)$ is achieved at σ^* for each $k \ge 0$ and

$$\mathrm{sep}^*(k; Q) = \mathrm{sep}_{\mathrm{id}}(k; Q) = P[V > k], \quad k \ge 0,$$

where

$$(6.7) \qquad \mathcal{L}(V) \;=\; \mathcal{L}(S \oplus T(\sigma^*))$$
$$(6.8) \qquad \phantom{\mathcal{L}(V)} \;=\; \mathcal{L}(S \oplus W_1 \oplus \cdots \oplus W_{n-2})$$

and the random variables in each sum are mutually independent. The random variable S is a time to stationarity for the random walk R started from any fixed state. The random variable $T(\sigma^*)$ is as in Theorem 6.1. For $j \in [n-2]$, the random variable W_j has the same distribution as a time to stationarity for the simple symmetric (non-circular) random walk of Corollary 4.1, with n there replaced by $v+1$, started in state 1. Furthermore,

$$V \sim \oplus_{j=1}^{m} \mathrm{Geom}(1 - \lambda_{j,m}) \oplus \oplus_{l=1}^{n-2} \oplus_{j=1}^{l} \mathrm{Geom}(1 - \lambda_{j,l+1}),$$

where

$$\lambda_{s,t} := 1 - 2p \left(1 - \cos\left(\frac{\pi s}{t} \right) \right).$$

Proof. We will first prove (b). After computing the expectation and variance of V, part (a) will follow by Chebychev's inequality.

Suppose that the MMTF chain is started at $\pi = \mathrm{id}$. Noting that $Q^\infty(\sigma^*) > 0$, the results of (b) up through (6.7) will follow from Theorem 6.1 once we show (i) σ_{n-1}^* achieves the separation $\mathrm{sep}_1(k; R)$ for every $k > 0$; (ii) $L(\sigma^*) = n - 1$; and (iii) σ^* maximizes $\mathcal{L}(T(\sigma))$ stochastically. In what follows we use several results from Diaconis and Fill (1990), which treats separation for several classes of Markov chains, including birth-and-death chains. In particular, (i) follows from the discussion in Section 4 of that paper (see especially the end of Example 4.46) where we have used the fact that the holding probability of the random walk satisfies $1/3 \leq 1 - 2p < 1$, a consequence of $0 < p \leq 1/4$, and (ii) is obvious.

To establish (iii) we use the result from the previous remark. Simplifying the notation there, let $W(v, \sigma)$ be a random variable whose distribution is the conditional distribution of $\widetilde{W}_{v+1}$ given $\widetilde{X}_{\widetilde{C}_{\to n}} = \sigma$. It suffices to show that σ^* maximizes $\mathcal{L}(W(v, \sigma))$ stochastically for $v \in [n - 2]$ among all $\sigma \in RC = \{\sigma \in S_n : Q^\infty(\sigma) > 0\}$.

For $v \in [n - 2]$ and $\sigma \in RC$, $R_{\sigma[v]}$ does not depend on σ. Write R_v for such $R_{\sigma[v]}$, and relabel the rows and columns sequentially with $1, \ldots, v$. Thus R_v is a tridiagonal $v \times v$ matrix with diagonal entries $1 - 2p$ and sub- and super-diagonal entries p. Arguing as in Feller (1968, Section XVI.3), we diagonalize R_v (omitting details) to obtain, for $k \geq 1$ and $i, j \in [v]$,

$$R_v^k(i, j) = \frac{2}{v + 1} \sum_{l=1}^{v} \sin\left(\frac{\pi l i}{v + 1}\right) \sin\left(\frac{\pi l j}{v + 1}\right) \lambda_{l, v+1}^k.$$

Since $|\lambda_{l, v+1}| < 1$ for all $l \in [v]$,

$$(I_v - R_v)^{-1}(i, j) = \sum_{k=0}^{\infty} R_v^k(i, j)$$

$$= \frac{1}{p(v + 1)} \sum_{l=1}^{v} \sin\left(\frac{\pi l i}{v+1}\right) \sin\left(\frac{\pi l j}{v+1}\right) \left(1 - \cos\left(\frac{\pi l}{v + 1}\right)\right)^{-1},$$

and hence all the ingredients of (6.6) can be calculated explicitly. For fixed v and $\sigma \in RC$,

$$P[W(v, \sigma) = w] = \begin{cases} \frac{R_v^{w-1}(1,1)}{(I_v - R_v)^{-1}(1,1)}, & \text{if } \sigma_{v+1} \sim \sigma_v \\ \frac{R_v^{w-1}(1,v)}{(I_v - R_v)^{-1}(1,v)}, & \text{if } \sigma_{v+1} \not\sim \sigma_v. \end{cases}$$

It is obvious from a directly probabilistic argument that

$$\mathcal{L}(W(v, \sigma)) = \mathcal{L}(\widetilde{W}_{v+1} | \widetilde{X}_{\widetilde{C}_{\to(v+1)}})$$

is stochastically larger in the case $\sigma_{v+1} \not\sim \sigma_v$ than in the case $\sigma_{v+1} \sim \sigma_v$. Since $\sigma_{v+1}^* \not\sim \sigma_v^*$ for all $v \in [n - 2]$, condition (iii) is met.

It is now straightforward to compute

$$P[W(v, \sigma^*) = w] = 2p \sum_{l=1}^{v} (-1)^{l-1} \sin^2\left(\frac{\pi l}{v + 1}\right) \lambda_{l, v+1}^{w-1}$$

(6.9)
$$= \sum_{l=1}^{v} \alpha(l)(1 - \lambda_{l, v+1}) \lambda_{l, v+1}^{w-1},$$

where

$$\alpha(l) := (-1)^{l-1}\left(1 + \cos\left(\frac{\pi l}{v+1}\right)\right), \quad l \in [v].$$

This distribution has a curious and useful interpretation. As shown in Section 4 of Diaconis and Fill (1990), if the *noncircular* walk (call it R'_{v+1}) described in Corollary 4.1, with n there replaced by $v + 1$, is started in state 1, then for every $k \geq 0$ the separation $\mathrm{sep}_1(k; R'_{v+1})$ is achieved at state $v + 1$ and so equals $1 - (v + 1)(R'_{v+1})^k(1, v + 1)$. Diagonalizing R'_{v+1}, one discovers

$$\mathrm{sep}_1(k; R'_{v+1}) = P[W(v, \sigma^*) > k], \quad k \geq 0,$$

using (6.9) for the right side. Thus $W(v, \sigma^*)$ is distributed as the time to stationarity for R'_{v+1} started at 1.

By (4.58) in Diaconis and Fill (1990) (in which the sum should be $\sum_{j=1}^d$) and the ensuing discussion, for every $k \geq 0$ the separation $\mathrm{sep}_1(k; R)$ equals

$$\mathrm{sep}_1(k; R) = 2\sum_{j=1}^m (-1)^{j-1} \cos\left(\frac{j\pi}{2m}\right)\left(1 - 2p\left(1 - \cos\left(\frac{j\pi}{m}\right)\right)\right)^k$$

and is achieved (uniquely) at state $m + 1$.

Again by the discussion in Diaconis and Fill (1990) (see especially Theorem 4.20), the probability generating function for S is given by

$$z \mapsto \prod_{j=1}^m \frac{(1 - \lambda_{j,m})z}{1 - \lambda_{j,m}z},$$

and W_l has probability generating function

$$z \mapsto \prod_{j=1}^l \frac{(1 - \lambda_{j,l+1})z}{1 - \lambda_{j,l+1}z}.$$

Since we assume $p \leq 1/4$, $\lambda_{s,t} \geq 0$ for all s and t and thus S and each W_l is distributed as the sum of independent geometric random variables, completing the proof of (b). When some of the λ's are negative, a distributional interpretation can be provided along the lines of Remark 4.22(c) in Diaconis and Fill (1990).

For (a) we note that $k(n, c) = E[V] + c\sqrt{\mathrm{Var}[V]}$ steps are necessary and sufficient by Chebychev's inequality. But

$$E[V] = \sum_{j=1}^m \frac{1}{1 - \lambda_{j,m}} + \sum_{l=1}^{n-2}\sum_{j=1}^l \frac{1}{1 - \lambda_{j,l+1}}$$

$$= \frac{1}{2p} \sum_{j=1}^{n/2} \frac{1}{1 - \cos(2\pi j/n)} + \frac{1}{2p} \sum_{l=1}^{n-2} \sum_{j=1}^{l} \frac{1}{1 - \cos(\pi j/(l+1))}$$

$$= \frac{n^3}{18p} + O(n^2)$$

and $\mathrm{Var}[V] = \sum_{j=1}^{m} \frac{\lambda_{j,m}}{(1 - \lambda_{j,m})^2} + \sum_{l=1}^{n-2} \sum_{j=1}^{l} \frac{\lambda_{j,l+1}}{(1 - \lambda_{j,l+1})^2}$

$$= \frac{1}{4p^2} \sum_{j=1}^{n/2} \frac{1 - 2p(1 - \cos(2\pi j/n))}{(1 - \cos(2\pi j/n))^2} + \frac{1}{4p^2} \sum_{l=1}^{n-2} \sum_{j=1}^{l} \frac{1 - 2p(1 - \cos(\pi j/(l+1)))}{(1 - \cos(\pi j/(l+1)))^2}$$

$$= \frac{n^5}{450p^2} + O(n^4),$$

using standard asymptotic analysis and the values $\sum_{k=1}^{\infty} k^{-2} = \pi^2/6$ and $\sum_{k=1}^{\infty} k^{-4} = \pi^4/90$ of the Riemann zeta function.   □

Remarks:

1. An alternative approach applying Markov's inequality to

$$\exp\left[\mathrm{const.} \times \frac{V - E[V]}{\sqrt{\mathrm{Var}[V]}}\right]$$

shows that $H(c)$ in the definition of "sufficient number of steps" can be taken to be of the form $H(c) = \alpha e^{-\beta c}$ where α and β are positive constants. Such exponential decay of H is sometimes required in the definition of "sufficient number of steps."

2. An even sharper result, at least asymptotically as $n \to \infty$, is that for each fixed $c \in \mathbb{R}$ and $k = \lfloor (18p)^{-1}n^3 + c(15\sqrt{2}\, p)^{-1}n^{5/2} \rfloor$,

$$\mathrm{sep}^*(k; Q) \to P[Z > c] \text{ as } n \to \infty,$$

where Z is a standard normal random variable. This can be shown using Liapounov's doubly indexed array version of the central limit theorem (e.g., Chung (1974, Section 7.2)) and a calculation that the sum of the fourth central moments of the component geometric random variables is of order n^9.

7. Total variation distance. In this section we use coupling to derive bounds on the total variation distance between MMTF and its stationary distribution.

Let $X = (X_n)$ and $Y = (Y_n)$ be two realizations of a Markov chain with transition matrix P. Suppose that the X-chain has an arbitrary initial distribution and the Y-chain is started (say) in stationarity. A *coupling time* T is a stopping time for the bivariate chain (X_n, Y_n) such that $X_n = Y_n$ for all $n \geq T$.

Our coupling for the MMTF chain is constructed in two steps: (i) First couple two copies of the request chain; (ii) then use the standard coupling for MTF, namely, wait until all but one of the records have been requested at least once. Thus the analysis for the MMTF chain reduces to an analysis of the much smaller request chain.

Let P^∞ denote the stationary distribution of the chain and let $P_i^k = P^k(i, \cdot)$ be the distribution after k steps when the chain is started in state i. The *total variation distance* between P_i^k and P^∞ is

$$(7.1) \quad d_i(k; P) := \|P_i^k - P^\infty\|_{TV} = \max_A |P_i^k(A) - P^\infty(A)| = \min_T P[T > k],$$

where the maximum in the third expression is over all events A and the minimum in the fourth expression is over all coupling times T for P_i^k and P^∞.

For the request chain X run in stationarity and $t \in \{0, 1, 2, \ldots\}$, let

$$C(t) := \inf\{k \geq 0 : \{X_t, X_{t+1}, \ldots, X_{t+k}\} = [n]\}$$

be the *ergodic cover time* for the chain X started at time t. Let $C := C(0)$. The following result is evident.

THEOREM 7.1. *For $i \in [n]$, let F_i be a fastest coupling time for the request chain started at state i. For $t \in \{0, 1, 2, \ldots\}$, let $C(t)$ be the ergodic cover time of the request chain started at time t. Let $\pi \in S_n$. Then $T_{\pi_1} := F_{\pi_1} + C(F_{\pi_1})$ is a coupling time for MMTF started in π. Furthermore, for any $\zeta \in [0, 1)$,*

$$
\begin{aligned}
d_\pi(k; Q) &\leq P[T_{\pi_1} > k] \\
&\leq P[F_{\pi_1} > \zeta k] + P[C > (1 - \zeta)k] \\
(7.2) \qquad &\leq d_{\pi_1}(\lfloor \zeta k \rfloor; R) + \frac{1}{(1 - \zeta)k} E[C].
\end{aligned}
$$

For our running random walk example, direct arguments give the following.

THEOREM 7.2. *Consider MMTF with request chain R as given in Corollary 4.1. Then cn^2 steps are necessary and sufficient for convergence to stationarity in total variation distance.*

Proof. The proof we sketch is based on Theorem 7.1 and well-known results for simple symmetric random walk on $[n]$.

For the lower bound, it takes cn^2 steps for the front record in the list, from any fixed initial state, to become uniform in total variation distance. Thus it takes at least cn^2 steps for the list as a whole to become stationary.

For the upper bound, cn^2 steps are sufficient for the request chain to become uniform in total variation distance. So, from any initial i, the fastest coupling F_i takes at most cn^2 steps. And cn^2 steps are always sufficient to cover $[n]$. The upper bound thus follows from Theorem 7.1. $\square$

Remark:

It is interesting to note the different behaviors for separation and total variation distance for this example. Convergence to stationarity in total variation distance exhibits no cutoff. On the other hand, convergence to stationarity in separation exhibits a cutoff and is slower by an order of magnitude.

Application of Theorem 7.1 (e.g., using the second and third inequalities in (7.2)) to more general request chains requires knowledge of the covering time C. Unfortunately, there are few results on the distribution of C for any but the most structured chains. There is a growing body of work which treats expected cover time, but these bounds are typically not sharp. A valuable source for such results is Broder and Karlin (1989).

Without going into details, this approach gives an upper bound in the case of the mixture model (2.1) introduced in Section 2 when R_0 is the matrix each of whose entries is $1/n$ and B is the request transition matrix discussed in Corollary 4.1. For fixed $\alpha \in (0, 1)$, order $n \log n$ steps are sufficient for convergence to stationarity in total variation distance. We conjecture that order $n \log n$ steps are also necessary.

REFERENCES

Aldous, D., and Diaconis, P. (1986). Shuffling cards and stopping times. *Amer. Math. Monthly* **93** 333–347.

Aldous, D., and Diaconis, P. (1987). Strong uniform times and finite random walks. *Adv. in Appl. Math.* **8** 69–97.

Bentley, J. L., and McGeoch, C. C. (1985). Amortized analyses of self-organizing sequential search heuristics. *Comm. ACM* **29** 404–411.

Bitner, J. R. (1979). Heuristics that dynamically organize data structures. *SIAM J. Comp.* **8** 82–110.

Broder, A. Z., and Karlin, A. R. (1989). Bounds on the cover time. *J. Theo. Prob.* **2** 101–120.

Chung, K. L. (1974). *A Course in Probability Theory*, 2nd edition. Academic Press, New York.

Diaconis, P. (1993). Notes on the weighted rearrangement process. Unpublished manuscript.

Diaconis, P., and Fill, J. A. (1990). Strong stationary times via a new form of duality. *Ann. Prob.* **18** 1483–1522.

Diaconis, P., Fill, J. A., and Pitman, J. (1992). Analysis of top to random shuffles. *Comb., Prob. and Comp.* **1** 135–155.

Dobrow, R. P. (1994). Markov chain analysis of some self-organizing schemes for lists and trees. Ph.D. dissertation, Department of Mathematical Sciences, The Johns Hopkins University.

Feller, W. (1968). *An Introduction to Probability Theory and Its Applications*, Vol. I. John Wiley & Sons, New York.

Fill, J. A. (1995). An exact formula for the move-to-front rule for self-organizing lists. *J. Theo. Prob.*, to appear.

Hendricks, W. J. (1989). *Self-organizing Markov Chains.* MITRE Corp., McLean, Va.

Hester, J. H., and Hirschberg, D. S. (1985). Self-organizing linear search. *Comp. Surveys* **17** 295–311.

Kapoor, S., and Reingold, E. M. (1991). Stochastic rearrangement rules for self-organizing data structures. *Algorithmica* **6** 278–291.

Knuth, D. (1973). *The Art of Computer Programming, Searching and Sorting,* Vol. III. Addison–Wesley, Reading, Mass.

Konnecker, L. K., and Varol, Y. L. (1981). A note on heuristics for dynamic organization of data structures. *Info. Proc. Lett.* **12** 213–216.

Lam, K., Leung, M.-Y., and Siu, M.-K. (1984). Self-organizing files with dependent accesses. *J. Appl. Prob.* **21** 343–359.

Phatarfod, R. M., and Dyte, D. (1993). The linear search problem with Markov dependent requests. Preprint.

Rivest, R. (1978). On self-organizing sequential search heuristics. *Comm. ACM* **19** 63–67.

Rodrigues, E.R. (1993) The performance of the move-to-front scheme under some particular forms of Markov requests. Preprint.

THE ASYMPTOTIC LOWER BOUND ON THE DIAGONAL RAMSEY NUMBERS: A CLOSER LOOK*

ANANT P. GODBOLE[†], DAPHNE E. SKIPPER[‡], AND RACHEL A. SUNLEY[§]

Abstract. The classical diagonal Ramsey number $R(k, k)$ is defined, as usual, to be the smallest integer n such that any two-coloring of the edges of the complete graph K_n on n vertices yields a monochromatic k-clique. It is well-known that $R(3,3) = 6$ and $R(4,4) = 18$; the values of $R(k,k)$ for $k \geq 5$, are, however, unknown. The Lovász local lemma has been used to provide the best-known asymptotic lower bound on $R(k,k)$, namely $\frac{\sqrt{2}}{e} k 2^{k/2} (1 + o(1))(k \to \infty)$. In other words, if one randomly two-colors the edges of K_n, then $\mathbf{P}(X_0 = 0) > 0$ if

$$n \leq \frac{\sqrt{2}}{e} k 2^{k/2}(1 + o(1)), \qquad (\clubsuit)$$

where X_0 denotes the number of monochromatic k-cliques. We use univariate and process versions of the Stein-Chen technique to show that much more is true under condition $(\clubsuit)$: in particular, we prove (i) that the distribution of X_0 can be approximated by that of a Poisson random variable, and that (ii) with X_j representing the number of k-cliques with exactly j edges of one color, the joint distribution of $(X_0, X_1, \ldots, X_b)$ can be approximated by a multidimensional Poisson vector with independent components provided that $b = o(k/\log k)$.

1. Introduction and statement of results. The diagonal Ramsey number $R(k, k)$ is defined to be the smallest integer n such that any two-coloring of the edges of the complete graph K_n on n vertices yields a monochromatic k-clique. It is well known that $R(3,3) = 6$ and $R(4,4) = 18$; the values of $R(k,k)$ are, however, unknown for $k \geq 5$. Erdős showed in 1947 that

$$(1.1) \qquad R(k,k) \geq \frac{1}{\sqrt{2}e} k 2^{k/2}(1 + o(1)), \quad (k \to \infty)$$

by using an elementary but far-reaching probabilistic argument: If we randomly color the edges of K_n red or blue, independently and with equal probability, and let X_0 denote the number of monochromatic k-cliques, then it is easy to verify that $n \leq \frac{1}{\sqrt{2}e} k 2^{k/2}(1 + o(1))$ implies that $\mathbf{E}(X_0) < 1$; this in turn forces $\mathbf{P}(X_0 = 0)$ to be positive. Erdős' result follows. A nice account of the asymptotic questions involved in the above argument may

* This research was partially supported by NSF Grant DMS-9200409. Part of the work reported herein was conducted at the Institüt für Angewandte Mathematik, Universität Zürich. Useful discussions with Andrew Barbour are gratefully acknowledged, as is the generous travel support derived from a Schweizerischer Nationalfonds Grant.

† Department of Mathematical Sciences, Michigan Technological University, Houghton, MI 49931, e-mail: anant@math.mtu.edu,

‡ Department of Mathematics, The University of the South, Sewanee, TN 37375, e-mail: skippde0@seraph1.sewanee.edu

§ Department of Mathematics, Amherst College, Amherst, MA 01002, e-mail: rasunley@ amherst.edu

be found in Alon and Spencer (1992), pp. 27-28. The method of alterations yields a slight improvement:

$$(1.2) \qquad R(k,k) \geq \frac{1}{e} k 2^{k/2} (1 + o(1)), \quad (k \to \infty),$$

while the Lovász local lemma has been employed to provide the best-known asymptotic lower bound

$$(1.3) \qquad R(k,k) \geq \frac{\sqrt{2}}{e} k 2^{k/2} (1 + o(1)), \quad (k \to \infty);$$

see Alon and Spencer (1992) for proofs of (1.2) and (1.3). The best known upper bound on $R(k,k)$, on the other hand, is of order $(4 - o(1))^k$: more specifically,

$$(1.4) \qquad R(k,k) \leq \frac{\binom{2k}{k}}{k^\alpha},$$

for an absolute constant α; a proof of this fact may be found in Graham, Rothschild and Spencer (1990). The problem of determining whether or not $\frac{R(k,k)}{k2^{k/2}} \to \infty$, and more generally, of computing $\lim_{k \to \infty} R(k,k)^{1/k}$ (assuming that the limit exists), remain tantalizingly open; recent reformulations of the problem can be found in Hildebrand (1993). In this paper, we use (1.3), which we rewrite as

$$(1.5) \qquad n \leq \frac{\sqrt{2}}{e} k 2^{k/2} (1 + o(1)) \Rightarrow \mathbf{P}(X_0 = 0) > 0,$$

as a starting point. It is clear from (1.5) that bounds on $R(k,k)$ yield statements about the mass placed by $\mathcal{L}(X_0)$ at the origin, where $\mathcal{L}(Z)$ denotes the distribution of the random variable Z; we wish, on the other hand, to find a distributional approximation for $\mathcal{L}(X_0)$ itself. Since

$$(1.6) \qquad X_0 = \sum_{j=1}^{\binom{n}{k}} I_j,$$

where the indicator variable I_j equals one iff the jth k-clique is monochromatic, and since $\pi_j = \mathbf{P}(I_j = 1) = 1/2^{\binom{k}{2}-1}$ is low (for large values of k), one expects that the distribution of X_0 would be close to that of $\mathrm{Po}(\lambda)$, the Poisson r.v. with mean $\lambda = \mathbf{E}(X_0)$. We make this statement precise by using the Stein-Chen method and a non-monotone coupling (the monograph by Barbour, Holst and Janson (1992) is the standard reference on the coupling approach to Stein-Chen approximation) to prove

THEOREM 1. *Consider a random two-coloring of the edges of the complete graph K_n on n vertices, with each edge being colored red or blue independently and with equal probability. Then, with X_0 denoting the number* *of monochromatic k-cliques,* d_{TV} *the usual variation distance and* $\text{Po}(\lambda)$ *the Poisson r.v. with mean* $\lambda = \mathbf{E}(X_0) = \binom{n}{k}/2^{\binom{k}{2}-1}$,

$$d_{TV}(\mathcal{L}(X_0), \text{Po}(\lambda)) \leq f(n, k)$$

where $f(n, k) \to 0$ *as* $k \to \infty$ *provided that* $n \leq \frac{\sqrt{2}}{e} k 2^{k/2}(1 + o(1))$.

Note that the condition that guarantees Poisson approximability is the same as in (1.5), with a different $o(1)$ function. Unfortunately, it is quite another matter to retrieve the best known lower bound on $R(k, k)$ from Theorem 1, and we were unsuccessful in our attempts to do so. On the other hand, we will show as a corollary that the bound (1.2) [which was obtained on using the method of alterations] *can* easily be recovered from Theorem 1.

We next ask a more complicated question: Given a random 2-coloring of the edges of K_n, what is the overall structure of the numbers of k-cliques that contain j edges of one color, where $j = 0, 1, \ldots, b$? Specifically, how does one approximate the joint distribution of the vector variable $(X_0, X_1, \ldots, X_b)$, where X_j denotes the number of k-cliques with exactly j red (or blue) edges? Our second theorem builds on other multivariate Poisson approximations obtained, e.g., by Arratia and Tavaré (1992) and Barbour, Godbole and Qian (1993), in the context of random permutations and random tournaments respectively:

THEOREM 2. *With the same notation as in the preceding paragraph, and with* $\lambda_j = \binom{n}{k}\binom{\binom{k}{2}}{j}/2^{\binom{k}{2}-1}$,

$$d_{TV}\left(\mathcal{L}(X_0, X_1, \ldots, X_b), \prod_{j=0}^{b} \text{Po}(\lambda_j)\right) \leq f(n, k, b),$$

where $f(n, k, b) \to 0$ $(k \to \infty)$ *provided that* $b = o(k/\log k)$ *and* $n \leq \frac{\sqrt{2}}{e} k 2^{k/2}(1 + o(1))$.

Note that Theorem 2 is valid, as was Theorem 1, under the same condition as in (1.5). In other words, if $n \leq \frac{\sqrt{2}}{e} k 2^{k/2}(1 + o(1))$, then not only is there a positive probability of having no monochromatic k-cliques of either color, but the distribution of the *number* of monochromatic k-cliques is approximately Poisson as well; furthermore, the "big picture" can be modeled by a Poisson process with *independent* components. We will show, in addition, that the upper bound $b = o(k/\log k)$ in Theorem 2 cannot be bettered if one wishes to preserve the condition $n \leq \frac{\sqrt{2}}{e} k 2^{k/2}(1+o(1))$; this fact is exhibited by showing that even a univariate Poisson approximation for $\mathcal{L}(X_b)$ fails if $b \neq o(k/\log k)$. On the other hand, it is easy to check that the event "a k-set has b edges of one color" is rare for $b = o(k^2)$, which suggests that $\mathcal{L}(X_b)$ might be approximately Poisson for a larger range of values of b; the failure of the Poisson paradigm when b exceeds $o(k/\log k)$ is, therefore, attributable to the fact that the dependencies between the indicators of these events grow significantly more severe with increasing b. It is quite conceivable that several kinds of clustering phenomena cause a compound Poisson approximation to be more appropriate for larger values of b. This matter is under investigation. We have chosen to use the coupling approach to Stein-Chen approximation to prove Theorems 1 and 2, but the local method (Arratia, Goldstein and Gordon (1989)) can be verified to give exactly the same results; this is not surprising since the two approaches usually yield equivalent bounds in the "dissociated" case, of which our set-up is an example. Finally, we mention that algorithmic questions have not been addressed in this paper; the recent work of Beck (1991) (for example) shows that several interesting questions concerning graph coloring have efficient deterministic or randomized algorithms associated with their solution.

2. Proofs

Proof of Theorem 1. Denote the clique $\{1, 2, \ldots, k\}$ by $\mathbf{1}$. Since the intersection patterns of each of the $N := \binom{n}{k}$ k-cliques are identical, Theorem 2.A in Barbour, Holst and Janson (1992) yields, on some simplification,

$$(2.1) \qquad d_{TV}(\mathcal{L}(X_0), \mathrm{Po}(\lambda)) \leq 2^{1-\binom{k}{2}} + \sum_{\beta \neq \mathbf{1}} \mathbf{P}(I_\beta \neq J_\beta),$$

for any choice of coupled variables $\{J_\beta\}_{\beta=1}^N$ that satisfy

$$(2.2) \qquad \mathcal{L}\{J_\beta : 1 \leq \beta \leq N\} = \mathcal{L}\{I_\beta, 1 \leq \beta \leq N | I_\mathbf{1} = 1\}.$$

The coupling is chosen as follows: If $\mathbf{1}$ is a monochromatic clique, we do nothing, and set $J_\beta = I_\beta$ for each β. If, however, $I_\mathbf{1} = 0$, we flip a fair coin with faces $\{R, B\}$, where $R =$'Red' and $B =$'Blue', and change each of the $\binom{k}{2}$ edges in $\mathbf{1}$ to the color determined by the coin toss, letting $J_\beta = 1$ if the βth k-set is monochromatic after the colors are changed as described. It is clear that (2.2) is satisfied. Now, $I_\beta \equiv J_\beta$ if $|\mathbf{1} \cap \beta| \leq 1$, so that we may restrict the sum in (2.1) to β's satisfying $2 \leq |\mathbf{1} \cap \beta| \leq k - 1$. There are $\binom{k}{r}\binom{n-k}{k-r}$ sets β such that $|\mathbf{1} \cap \beta| = r$, and for each $r \in \{2, 3, \ldots, k-1\}$,

$$\mathbf{P}(I_\beta = 1, J_\beta = 0)$$

$$(2.3) \quad \leq \mathbf{P}(\beta \text{ was monochromatic; the coin flip was of the other color})$$

$$= 2^{-\binom{k}{2}},$$

so that

$$(2.4) \qquad \begin{aligned} \sum_{\beta \neq \mathbf{1}} \mathbf{P}(I_\beta = 1, J_\beta = 0) &\leq 2^{-\binom{k}{2}} \sum_{r=2}^{k-1} \binom{k}{r}\binom{n-k}{k-r} \\ &\leq 2^{-\binom{k}{2}} \binom{k}{2}\binom{n}{k-2} \\ &\leq 2^{-\binom{k}{2}} \frac{k^2}{2} [\frac{ne}{k-2}]^{k-2} \end{aligned}$$

Next consider $\mathbf{P}(I_\beta = 0, J_\beta = 1)$: Assume that $2 \le r := |\mathbf{1} \cap \beta| \le k - 1$. It is clear, that for a non-monochrome clique to change into one that is monochrome, (i) each of the $\binom{k-r}{2}$ edges in $\beta \setminus \mathbf{1}$ *as well as* (ii) each of the $r(k - r)$ edges connecting points in $\beta \cap \mathbf{1}$ and $\beta \setminus \mathbf{1}$, must have been of the same color to begin with. The coin toss, must, moreover, have yielded the same color. It follows that for $\beta \cap \mathbf{1} = r$,

$$(2.5) \qquad \mathbf{P}(I_\beta = 0, J_\beta = 1) \le 2^{-\binom{k-r}{2} - r(k-r)},$$

and thus, using (2.4), that

$$
\begin{aligned}
(2.6) \quad &\sum_{\beta \ne \mathbf{1}} \mathbf{P}(I_\beta \ne J_\beta) \\
&\le 2^{-\binom{k}{2}} \frac{k^2}{2} \left[\frac{ne}{k-2} \right]^{k-2} + \sum_{r=2}^{k-1} \binom{k}{r} \binom{n-k}{k-r} 2^{-\binom{k-r}{2} - r(k-r)} \\
&= 2^{-\binom{k}{2}} \frac{k^2}{2} \left[\frac{ne}{k-2} \right]^{k-2} + \sum_{s=1}^{k-2} \binom{k}{s} \binom{n-k}{s} 2^{\frac{s(s+1)}{2} - ks} \\
&= 2^{-\binom{k}{2}} \frac{k^2}{2} \left[\frac{ne}{k-2} \right]^{k-2} + \sum_{s=1}^{k-2} a_s, \quad \text{say.}
\end{aligned}
$$

Now a_s can be shown, on employing a tedious but straightforward argument, to be decreasing and then increasing, so that we may bound the series in (2.6) by $ka_1 + ka_{k-2}$ to obtain

$$
\begin{aligned}
(2.7) \quad &\sum_{\beta \ne \mathbf{1}} \mathbf{P}(I_\beta \ne J_\beta) \\
&\le 2^{-\binom{k}{2}} \frac{k^2}{2} \left[\frac{ne}{k-2} \right]^{k-2} + \frac{k^2 n}{2^{k-1}} + \frac{k \binom{k}{2} \binom{n-k}{k-2}}{2^{\binom{k}{2}-1}} \\
&\le \frac{k^2 n}{2^{k-1}} + \left\{ k^3 + \frac{k^2}{2} \right\} \frac{\left[\frac{ne}{k-2} \right]^{k-2}}{2^{\binom{k}{2}}} \\
&\le \frac{k^2 n}{2^{k-1}} + \frac{k^3 \left[\frac{ne}{k-2} \right]^{k-2}}{2^{\binom{k}{2}-1}},
\end{aligned}
$$

which, together with (2.1) yields

$$(2.8) \quad d_{TV}(\mathcal{L}(X_0), \mathrm{Po}(\lambda)) \le 2^{1-\binom{k}{2}} + \frac{k^2 n}{2^{k-1}} + \frac{k^3 \left[\frac{ne}{k-2} \right]^{k-2}}{2^{\binom{k}{2}-1}} := f(n, k)$$

Now the first term on the right hand side of (2.8) is small if k is large, while the second term is negligible whenever $n = o(2^{k-1}/k^2)$, so that the last term clearly determines the magnitude of the error bound $f(n, k)$. If

$$(2.9) \qquad n \le \frac{\sqrt{2}}{e} k 2^{\frac{k}{2} - \frac{(3+\epsilon) \log_2 k}{k}} \qquad (\epsilon > 0),$$

i.e., if $n \le \frac{\sqrt{2}}{e} k 2^{k/2}(1 + o(1))$, then

(2.10)
$$\frac{k^3[\frac{ne}{k-2}]^{k-2}}{2^{\binom{k}{2}-1}} \le \frac{k^3[\frac{k}{k-2} 2^{\frac{k}{2}+\frac{1}{2}-\frac{(3+\epsilon)\log_2 k}{k}}]^{k-2}}{2^{\binom{k}{2}-1}}$$

$$\le \frac{2e^2 k^3}{k^{(3+\epsilon)(\frac{k-2}{k})}} = o(1) \quad (k \to \infty),$$

establishing Theorem 1. $\square$

We now attempt to use Theorem 1 to retrieve the lower bounds (1.1) through (1.3) on $R(k,k)$: By (2.8), $P(X_0 = 0) \ge e^{-\lambda} - f(n,k)$. It may easily be verified that $e^{-\lambda} > f(n,k)$ when $n \le \frac{1}{\sqrt{2e}} k 2^{k/2}(1 + o(1))$; this yields (1.1). In an attempt to do better, we proceed as follows: Since

(2.11) $$\mathbf{P}(X_0 \le \lambda) \ge \mathbf{P}(\mathrm{Po}(\lambda) \le \lambda) - f(n,k) \sim \frac{1}{2} - f(n,k),$$

we consider the inequality $f(n,k) < 1/2$, which would yield a positive probability of there existing a random two-coloring of the edges of K_n with *at most* λ monochromatic k-cliques. If we now delete from the graph one vertex from each such monochromatic k-set, we would have removed at most λ vertices, and the residual graph would have no monochrome cliques. It would follow that $R(k,k) \ge n - \lambda$. To this end, we note that the first and second components of the right hand side of (2.8) are smaller than $1/8$ if, respectively, k is large enough and $n \le 2^{k-4}/k^2$. The third term again dominates; it is smaller than $1/4$ if

(2.12) $$8k^3[\frac{ne}{k-2}]^{k-2} < 2^{\binom{k}{2}}.$$

Now, if (2.9) holds, then (2.12) may be verified to be true provided that

(2.13) $$\frac{8e^2 2^{\binom{k}{2}}}{k^{\frac{(3+\epsilon)(k-2)}{k}-3}} < 2^{\binom{k}{2}},$$

which is true if k is large enough. Thus $f(n,k) < 1/2$ if

(2.14) $$n \le \frac{\sqrt{2}}{e} k 2^{\frac{k}{2}-\frac{(3+\epsilon)\log_2 k}{k}} = N_0, \quad (\text{say}).$$

Unfortunately, for $n = N_0$, λ is of order $(\frac{ne}{k})^k 2^{1-\binom{k}{2}} \approx 2^k$, so that the lower bound $R(k,k) > N_0 - \lambda_{N_0}$ is useless. We work, therefore, with the choice $N_1 = \frac{1}{e} k 2^{k/2}(1 + o(1))$; this yields

(2.15) $$\lambda_{N_1} \le (\frac{ne}{k})^k 2^{1-\binom{k}{2}} = 2 \cdot 2^{k/2},$$

so that

$$R(k, k) > N_1 - \lambda_{N_1}$$

(2.16)
$$\geq \tfrac{1}{e} k 2^{k/2} (1 + o(1)) - 2 \cdot 2^{k/2}$$

$$= \tfrac{1}{e} k 2^{k/2} (1 + o(1))$$

In conclusion, therefore, we are able to rederive (1.1) and (1.2), but not (1.3), from Theorem 1. In retrospect, this is not surprising, since the latter result provides a total variation comparison between $\mathcal{L}(X_0)$ and $\mathrm{Po}(\lambda)$, whereas the lower bound (1.3) was obtained on using a powerful method (the local lemma) specifically designed to estimate $\mathbf{P}(X_0 = 0)$.

Proof of Theorem 2. Our starting point will be Theorem 10.K in Barbour, Holst and Janson (1992): Consider a collection $\{I_\gamma\}_{\gamma \in \Gamma}$ of indicator r.v.'s. Assume that a partition $\Gamma = \cup_{\alpha=0}^b \Gamma_\alpha$ of the index set Γ is given, and that the random variables $X_\alpha = \sum_{\gamma \in \Gamma_\alpha} I_\gamma, 0 \leq \alpha \leq b$ are of interest. Set $\lambda_\alpha = \mathbf{E}(X_\alpha)$. If for each γ, there exists a coupling $\{J_{\delta\gamma}\}_{\delta \in \Gamma}$ such that

(2.17)
$$\mathcal{L}(\{J_{\delta\gamma}\}, \delta \in \Gamma) = \mathcal{L}(I_\delta, \delta \in \Gamma | I_\gamma = 1)$$

then, with $\pi_\gamma = \mathbf{P}(I_\gamma = 1)$,

$$d_{TV}(\mathcal{L}(\{X_\alpha\}_{\alpha=0}^b), \textstyle\prod_{\alpha=0}^b \mathrm{Po}(\lambda_\alpha))$$

(2.18)
$$\leq \frac{1 + 2\log^+(e \min \lambda_\alpha)}{e \min \lambda_\alpha} \sum_\gamma \{\pi_\gamma^2 + \sum_{\delta \neq \gamma} [\pi_\gamma \mathbf{P}(J_{\delta\gamma} \neq I_\delta)]\};$$

(2.18) significantly improves on, e.g., Theorem 10.J, found on the preceding page of Barbour, Holst and Janson (1992). While it is generally believed that the multiplier $[1 + 2\log^+(e \min \lambda_\alpha)]/e \min \lambda_\alpha$ can be improved upon, *it will turn out that (2.18) will be quite adequate for the job of proving Theorem 2, whereas the weaker Theorem 10.J would not have been.* We choose the two-dimensional index set $\Gamma = \{(i, \alpha) : 1 \leq i \leq \binom{n}{k}; 0 \leq \alpha \leq b\}$ and set $I_{i,\alpha} = 1$ iff the ith k-clique has exactly α blue (or red) edges after the random two-coloring is carried out. It is clear that for each $\alpha = 0, 1, \ldots, b$, $X_\alpha = \sum_{i=1}^{\binom{n}{k}} I_{i,\alpha}$, that $\pi_{i,\alpha} = \binom{\binom{k}{2}}{\alpha} 2^{1 - \binom{k}{2}}$ for each i, and thus that $\lambda_\alpha = \binom{n}{k}\binom{\binom{k}{2}}{\alpha} 2^{1 - \binom{k}{2}}$. A little reflection shows that the following coupling satisfies (2.17): If the ith k-clique has exactly α blue (or red) edges, i.e., if $I_{i,\alpha} = 1$, then we do nothing - and let $J_{j,\beta}(= J_{(j,\beta),(i,\alpha)}) = I_{j,\beta}$ for each j and β. If, however, the ith k-clique doesn't have α edges of either color, we consider the color with the fewer number of edges (a coin flip could break the tie in the unlikely event that there are $\binom{k}{2}/2$ edges of each color). This color is equally likely to be red or blue; we assume, without loss of generality, that it is blue. If the number of blue edges is less than α, we randomly replace red edges by blue ones, so as to bring the number of blue edges up to par;

if, on the other hand, there are more than α blue edges in the ith clique, we recolor (randomly) an appropriate number of these edges red, so as to yield α blue edges as required. Finally, we set $J_{j,\beta}(= J_{(j,\beta),(i,\alpha)}) = 1$ if the jth k-clique has exactly β edges of one color *after* this change is instituted. Denote $[1 + 2\log^+(e\min\lambda_\alpha)]/e\min\lambda_\alpha = [1 + 2\log^+(e\lambda_0)]/e\lambda_0$ by $\phi(n,k)$. Since each k-clique has the same intersection pattern as the others, (2.18) reduces to

$$d_{TV}(\mathcal{L}(\{X_\alpha\}_{\alpha=0}^b), \prod_{\alpha=0}^b \text{Po}(\lambda_\alpha))$$

$$\leq \phi(n,k) \sum_{i,\alpha} \left\{ \mathbf{P}^2(I_{i,\alpha} = 1) + \sum_{j,\beta \neq i,\alpha} [\mathbf{P}(I_{i,\alpha} = 1)\mathbf{P}(J_{j,\beta} \neq I_{j,\beta})] \right\}$$

$$= \phi(n,k) \sum_{i=1}^{\binom{n}{k}} \sum_{\alpha=0}^b \left\{ \mathbf{P}^2(I_{i,\alpha} = 1) + \sum_{j,\beta \neq i,\alpha} [\mathbf{P}(I_{i,\alpha} = 1)\mathbf{P}(J_{j,\beta} \neq I_{j,\beta})] \right\}$$

$$= \phi(n,k)\binom{n}{k} \sum_{\alpha=0}^b \left\{ \mathbf{P}^2(I_{\mathbf{1},\alpha} = 1) + \sum_{j,\beta \neq \mathbf{1},\alpha} [\mathbf{P}(I_{\mathbf{1},\alpha} = 1)\mathbf{P}(J_{j,\beta} \neq I_{j,\beta})] \right\}$$

$$= \phi(n,k)\binom{n}{k} \sum_{\alpha=0}^b \left\{ \binom{\binom{k}{2}}{\alpha}^2 2^{2-k(k-1)} + \sum_{\beta \neq \alpha} \mathbf{P}(I_{\mathbf{1},\alpha} = 1)\mathbf{P}(J_{\mathbf{1},\beta} \neq I_{\mathbf{1},\beta}) \right.$$

$$\left. + \sum_{j \neq \mathbf{1}} \sum_{\beta=0}^b \mathbf{P}(I_{\mathbf{1},\alpha} = 1)\mathbf{P}(J_{j,\beta} \neq I_{j,\beta}) \right\}$$

$$= T_1 + T_2 + T_3, \quad \text{say,}$$

(2.19)
where $\mathbf{1}$ denotes, as usual, the clique $\{1, 2, \ldots, k\}$. Consider T_2. Since $\mathbf{P}(I_{\mathbf{1},\beta} \neq J_{\mathbf{1},\beta}) = \mathbf{P}(I_{\mathbf{1},\beta} = 1) = \binom{\binom{k}{2}}{\beta}2^{1-\binom{k}{2}}$, (2.19) reduces to

$$d_{TV}(\mathcal{L}(\{X_\alpha\}_{\alpha=0}^b), \prod_{\alpha=0}^b \text{Po}(\lambda_\alpha))$$

$$\leq \phi(n,k)\binom{n}{k} \sum_{\alpha=0}^b \left\{ \binom{\binom{k}{2}}{\alpha}^2 2^{2-k(k-1)} + \sum_{\beta \neq \alpha} \binom{\binom{k}{2}}{\alpha}\binom{\binom{k}{2}}{\beta}2^{2-k(k-1)} \right.$$

$$\left. + \sum_{j \neq \mathbf{1}} \sum_{\beta=0}^b \mathbf{P}(I_{\mathbf{1},\alpha} = 1)\mathbf{P}(J_{j,\beta} \neq I_{j,\beta}) \right\}$$

(2.20)
We now focus on the last sum T_3 in (2.20). Letting $l :=$ the number of edges of the "recessive" color (i.e., the color with the fewer number of edges) in the clique $\mathbf{1}$, we consider three cases

(a) $I_{j,\beta} = 1$, $J_{j,\beta} = 0$
(b) $0 \leq l \leq \alpha - 1$, $I_{j,\beta} = 0$, $J_{j,\beta} = 1$

and

(c) $\alpha + 1 \leq l \leq \binom{k}{2}/2$, $I_{j,\beta} = 0$, $J_{j,\beta} = 1$

while reminding the reader that $J_{j,\beta}$ is shorthand for $J_{(j,\beta),(1,\alpha)}$. We start with case (a):

$$\phi(n,k)\binom{n}{k}\sum_{\alpha=0}^{b}\sum_{j\neq 1}^{b}\sum_{\beta=0}^{b}\mathbf{P}(I_{1,\alpha}=1)\mathbf{P}(J_{j,\beta}=0,I_{j,\beta}=1)$$

$$\leq \frac{1+2\log^+(e\lambda_0)}{e2^{1-\binom{k}{2}}}\sum_{\alpha=0}^{b}\mathbf{P}(I_{1,\alpha}=1)\sum_{j,\beta}\mathbf{P}(J_{j,\beta}=0,I_{j,\beta}=1)$$

$$\leq \frac{1+2\log^+(e\lambda_0)}{e2^{1-\binom{k}{2}}}\sum_{\alpha=0}^{b}\pi_\alpha\sum_{r=2}^{k-1}\binom{k}{r}\binom{n-k}{k-r}\sum_{\beta}\mathbf{P}(I_{j,\beta}=1)$$

(2.21)

$$\leq \frac{1+2\log^+(e\lambda_0)}{e2^{1-\binom{k}{2}}}b^2\pi_b^2\sum_{r=2}^{k-1}\binom{k}{r}\binom{n-k}{k-r}$$

$$\leq \frac{1+2\log^+(e\lambda_0)}{e2^{1-\binom{k}{2}}}b^2\pi_b^2\binom{k}{2}\binom{n}{k-2}$$

$$\leq \frac{1+2\log^+(e\lambda_0)}{e}2^{\binom{k}{2}-1}k^2\binom{\binom{k}{2}}{b}^2\frac{1}{2^{k(k-1)-2}}\frac{k^2}{2}[\frac{ne}{k-2}]^{k-2}\quad(b\leq k)$$

$$\leq \frac{1+2\log^+(e\lambda_0)}{e}\frac{k^4}{2^{\binom{k}{2}}}(\frac{k^2e}{2b})^{2b}[\frac{ne}{k-2}]^{k-2},$$

where $\pi_\alpha=\pi_{1,\alpha}$. We next consider the $\log^+$ multiplier in (2.21); assuming, as we may, that $n\leq k2^{k/2}$,

$$(2.22)\qquad\lambda_0=\frac{\binom{n}{k}}{2^{\binom{k}{2}-1}}\leq(\frac{ne}{k})^k\frac{1}{2^{\binom{k}{2}-1}}\leq 2e^k2^{k/2},$$

so that $\log^+(e\lambda_0)\leq Ak$ for some A. (2.21) thus yields

$$\phi(n,k)\binom{n}{k}\sum_{\alpha=0}^{b}\sum_{j\neq 1}\sum_{\beta=0}^{b}\mathbf{P}(I_{1,\alpha}=1)\mathbf{P}(J_{j,\beta}=0,I_{j,\beta}=1)$$

(2.23)

$$\leq\frac{Ak^5}{2^{\binom{k}{2}}}(\frac{k^2e}{2b})^{2b}[\frac{ne}{k-2}]^{k-2}.$$

Now, if $n\leq\frac{\sqrt{2}}{e}k2^{\frac{k}{2}-\frac{(5+\epsilon)\log_2 k}{k}}/(1+o_1(1))=\frac{\sqrt{2}}{e}k2^{\frac{k}{2}}(1+o(1))$ for a $o_1(1)$ function to be specified later, then it is easy to see that the contribution of case (a), to the third term in (2.20) is at most T_{3a}, where

$$T_{3a}\leq\frac{Ak^5}{2^{\binom{k}{2}}}(\frac{k^2e}{2b})^{2b}(\frac{k2^{\frac{k}{2}+\frac{1}{2}-\frac{(5+\epsilon)\log_2 k}{k}}}{(k-2)(1+o_1(1))})^{k-2}$$

(2.24)

$$\leq Bk^5(\frac{k^2e}{2b})^{2b}\frac{2^{-(5+\epsilon)(\frac{k-2}{k})\log_2 k}}{(1+o_1(1))^k}$$

If $b=o(k/\log k)$, i.e., if $b=k/[\log k\, g(k)]$, where $g(k)\to\infty$, then

$$(2.25)\quad(\frac{k^2e}{2b})^{2b}=(\frac{ke\log k\, g(k)}{2})^{\frac{2k}{\log k\, g(k)}}=(1+o_1(1))^k,\quad\text{say.}$$

Thus, if $o_1(1)$ is chosen as in (2.25), then (2.24) reveals that

$$(2.26) \qquad\qquad T_{3a} \leq \frac{Bk^5}{k^{(5+\epsilon)(\frac{k-2}{k})}} \to 0.$$

Consider next the contribution T_{3b} of case (b) to the third term in (2.20). We remind ourselves that there are no more than $\alpha - 1$ edges of the recessive color in the clique $\mathbf{1}$. Observe that

$$T_{3b} = \phi(n,k)\binom{n}{k} \sum_{\alpha=0}^{b} \sum_{j \neq \mathbf{1}} \sum_{\beta=0}^{b} \mathbf{P}(I_{\mathbf{1},\alpha}=1)\mathbf{P}(J_{j,\beta}=1, I_{j,\beta}=0)$$

$$= \phi(n,k)\binom{n}{k} \sum_{\alpha=0}^{b} \binom{\binom{k}{2}}{\alpha} \frac{1}{2^{\binom{k}{2}-1}} \sum_{r} \binom{k}{r}\binom{n-k}{k-r} \sum_{\beta} \sum_{l=0}^{\alpha-1} \sum_{\zeta} \frac{\binom{\binom{k}{2}-\binom{r}{2}}{\binom{k}{2}-l-\zeta}\binom{\binom{r}{2}}{\zeta}}{2^{\binom{k}{2}-1}}$$

$$\cdot \sum_{\gamma} \frac{\binom{\zeta}{\gamma}\binom{\binom{k}{2}-l-\zeta}{\alpha-\gamma-l}}{\binom{\binom{k}{2}-l}{\alpha-l}} \frac{1}{2^{\binom{k}{2}-\binom{r}{2}}} \left\{ \binom{\binom{k}{2}-\binom{r}{2}}{\beta-\zeta+\gamma} + \binom{\binom{k}{2}-\binom{r}{2}}{\beta-\binom{r}{2}+\zeta-\gamma} \right\},$$

(2.27)

where $r :=$ the number of vertices in $\mathbf{1} \cap j$;
$l \leq k-1 :=$ the number of edges of the recessive color (=blue, say) in the k-clique $\mathbf{1}$;
$\zeta :=$ the number of edges, in $\mathbf{1} \cap j$, of the color that we are changing (red, in this case);
and
$\gamma :=$ the number of red edges, in $\mathbf{1} \cap j$, that are actually changed.
The first component $\binom{\binom{k}{2}-\binom{r}{2}}{\binom{k}{2}-l-\zeta}\binom{\binom{r}{2}}{\zeta}/2^{\binom{k}{2}-1}$ of the last part of (2.27) represents the probability of there being ζ red edges in $\mathbf{1} \cap j$, and a total of $\binom{k}{2} - l$ red edges in $\mathbf{1}$. The second piece $\binom{\zeta}{\gamma}\binom{\binom{k}{2}-l-\zeta}{\alpha-\gamma-l}/\binom{\binom{k}{2}-l}{\alpha-l}$ represents the probability that the coupling is implemented in a way that corresponds to γ red edges in $\mathbf{1} \cap j$ being changed to blue (note that a total of $\alpha - l$ red edges change color). Finally, the last portion $\left\{ \binom{\binom{k}{2}-\binom{r}{2}}{\beta-\zeta+\gamma} + \binom{\binom{k}{2}-\binom{r}{2}}{\beta-\binom{r}{2}+\zeta-\gamma} \right\}/2^{\binom{k}{2}-\binom{r}{2}}$ of (2.27) is the probability that the other edges in the set j have just the right number of edges *initially*, so as to lead to β red or β blue edges *after* the coupling is carried out; note that the 'other' edges in j are of two types: ones that connect two vertices in $j \setminus \mathbf{1}$, and those that connect vertices in $j \setminus \mathbf{1}$ and $\mathbf{1} \cap j$. We next bound the complicated right hand side of (2.27)

in the following strategically chosen fashion:

T_{3b}

$$\leq \phi(n,k)\binom{n}{k} \sum_{\alpha=0}^{b} \binom{\binom{k}{2}}{\alpha} \frac{1}{2^{\binom{k}{2}-1}} \sum_{r} \binom{k}{r}\binom{n-k}{k-r} \sum_{\beta} \sum_{l=0}^{\alpha-1} \sum_{\zeta} \frac{\binom{\binom{k}{2}-\binom{r}{2}}{\binom{k}{2}-l-\zeta}\binom{\binom{r}{2}}{\zeta}}{2^{\binom{k}{2}-1}}$$

$$\leq \phi(n,k)\binom{n}{k} \sum_{\alpha=0}^{b} \binom{\binom{k}{2}}{\alpha} \frac{1}{2^{\binom{k}{2}-1}} \sum_{r} \binom{k}{r}\binom{n-k}{k-r} \sum_{\beta} \sum_{l=0}^{\alpha-1} \frac{\binom{\binom{k}{2}}{i}}{2^{\binom{k}{2}-1}}$$

$$\leq \phi(n,k)\binom{n}{k}(b\pi_b)\binom{k}{2}\binom{n}{k-2}b^2\pi_b.$$

(2.28)

The right hand side of (2.28) can be shown to go to zero if $n \leq \frac{\sqrt{2}}{e}k2^{k/2}(1+o(1))$ and $b = o(k/\log k)$, in exactly the same way as the T_{3a} term. The extra 'b' in the upper bound (2.28) can be taken care of by choosing the $o(1)$ function differently; specifically, we take $n \leq \frac{\sqrt{2}}{e}k2^{\frac{k}{2}-\frac{(6+\epsilon)\log_2 k}{k}}/(1+o_1(1))$, where $o_1(1)$ is as before. Next, we consider the term T_{3c} that bounds the contribution of case (c) to the third term in (2.20). There are $l > \alpha$ edges of the recessive color (blue) in $\mathbf{1}$, so that we must change $l - \alpha$ blue edges to red. As in (2.27),

$T_{3c} \leq$

$$\phi(n,k)\binom{n}{k} \sum_{\alpha=0}^{b} \binom{\binom{k}{2}}{\alpha} \frac{1}{2^{\binom{k}{2}-1}} \sum_{r} \binom{k}{r}\binom{n-k}{k-r}$$

(2.29)
$$\cdot \sum_{\beta} \sum_{l} \sum_{\zeta} \binom{\binom{r}{2}}{\zeta}\binom{\binom{k}{2}-\binom{r}{2}}{l-\zeta} \frac{1}{2^{\binom{k}{2}-1}}$$

$$\cdot \sum_{\gamma} \frac{\binom{\zeta}{\gamma}\binom{l-\zeta}{l-\alpha-\gamma}}{\binom{l}{l-\alpha}} \frac{1}{2^{\binom{k}{2}-\binom{r}{2}}} \left\{ \binom{\binom{k}{2}-\binom{r}{2}}{\beta-\zeta+\gamma} + \binom{\binom{k}{2}-\binom{r}{2}}{\beta-\binom{r}{2}+\zeta-\gamma} \right\}$$

Since $\zeta \geq \gamma$, $\beta - \zeta + \gamma \leq \beta \leq b \ll k/\log k \ll \binom{k}{2} - \binom{r}{2}$, so that $\binom{\binom{k}{2}-\binom{r}{2}}{\beta-\zeta+\gamma} \leq \binom{\binom{k}{2}-\binom{r}{2}}{b} \leq \binom{\binom{k}{2}}{b}$. It may similarly be seen that $\beta - \binom{r}{2} + \zeta - \gamma \leq \beta$; in other words, $\binom{\binom{k}{2}-\binom{r}{2}}{\beta-\binom{r}{2}+\zeta-\gamma} \leq \binom{\binom{k}{2}-\binom{r}{2}}{b} \leq \binom{\binom{k}{2}}{b}$. (2.29) now yields

$$T_{3c} \leq 2b\phi(n,k)\binom{n}{k}\binom{\binom{k}{2}}{b} \sum_{\alpha=0}^{b} \binom{\binom{k}{2}}{\alpha} \frac{1}{2^{\binom{k}{2}-1}} \sum_{r} \frac{\binom{k}{r}\binom{n-k}{k-r}}{2^{\binom{k}{2}-\binom{r}{2}}}$$

(2.30)
$$\leq Akb^2\binom{\binom{k}{2}}{b}^2 \sum_{r} \frac{\binom{k}{r}\binom{n-k}{k-r}}{2^{\binom{k}{2}-\binom{r}{2}}}$$

$$\leq Akb^2\binom{\binom{k}{2}}{b}^2 \left\{ \frac{k^2 n}{2^{k-1}} + \frac{k^3[\frac{ne}{k-2}]^{k-2}}{2^{\binom{k}{2}-1}} \right\}$$

as in the proof of Theorem 1. The right hand side of (2.30) may be checked, as before, to tend to zero provided that $b = o(k/\log k)$ and $n \leq \frac{\sqrt{2}}{e} k 2^{\frac{k}{2} - \frac{(6+\epsilon)\log_2 k}{k}} / (1 + o_1(1))$. Finally, we consider the T_1 and T_2 terms from (2.19) and (2.20). T_2 dominates the two, and is evidently bounded by

$$(2.31) \qquad \phi(n,k)\binom{n}{k}b^2\left(\binom{\binom{k}{2}}{b}\right)^2 2^{2-k(k-1)} \leq \frac{k^3}{2^{\binom{k}{2}-1}}\left(\frac{k^2 e}{2b}\right)^{2b} \to 0;$$

this proves Theorem 2. $\square$

The proof of Theorem 2 involved several steps that might cause concern. In particular, the analysis leading up to (2.28) and (2.30) might be considered to be rather cavalier, with large portions of the exact expression in question being sacrificed. We show next that this is probably *not* the case, and that Theorem 2 is the best possible - in the sense that it is unlikely that the condition $b = o(k/\log k)$ can be bettered. We exhibit this fact by showing that even a *univariate* Poisson approximation for $\mathcal{L}(X_b)$ is invalid if $b \not\ll k/\log k$. For variety, we prove this optimality result by using a result of Barbour and Eagleson (reported as Corollary 2.C.5 in Barbour, Holst and Janson (1992)), invoking, in addition, the fact that the bounds given there are known to often be of the best order possible:

PROPOSITION. *With the same notation as in the rest of the paper, and assuming that $n \leq \frac{\sqrt{2}}{e} k 2^{k/2} (1 + o(1))$,*

$$d_{TV}(\mathcal{L}(X_b), \mathrm{Po}(\lambda_b)) \leq f(n,k,b)$$

where $f(n,k,b) \not\to 0$ as $k \to \infty$ if $b \not\ll k/\log k$.

Proof. Corollary 2.C.5 in Barbour, Holst and Janson (1992) yields, with $I_j = 1$ if the jth clique has b edges of one color,

$$d_{TV}(\mathcal{L}(X_b), \mathrm{Po}(\lambda_b)) \leq \frac{1 - e^{-\lambda_b}}{\lambda_b} \bullet$$

$$\sum_{j=1}^{\binom{n}{k}} \left\{ \mathbf{P}^2(I_j = 1) + \sum_r \binom{k}{r}\binom{n-k}{k-r}[\mathbf{P}(I_j = 1)\mathbf{P}(I_i = 1) + \mathbf{P}(I_i I_j = 1)] \right\}$$

$$\leq \frac{\binom{n}{k}}{\lambda_b}\left\{ \left(\binom{\binom{k}{2}}{b}\right)^2 2^{2-k(k-1)} + \right.$$

$$\left. \sum_r \binom{k}{r}\binom{n-k}{k-r}[\left(\binom{\binom{k}{2}}{b}\right)^2 2^{2-k(k-1)} + \mathbf{P}(I_i I_1 = 1)] \right\}$$

(2.32)

$$\leq \binom{\binom{k}{2}}{b}2^{1-\binom{k}{2}}\binom{k}{2}\binom{n}{k-2} + \frac{2^{\binom{k}{2}-1}}{\binom{\binom{k}{2}}{b}}\sum_r\binom{k}{r}\binom{n-k}{k-r}\mathbf{P}(I_iI_\mathbf{1}=1)$$

$$\leq \left(\frac{k^2e}{2b}\right)^b\frac{1}{2^{\binom{k}{2}-1}}\frac{k^2}{2}\left[\frac{ne}{k-2}\right]^{k-2}+$$

$$\frac{2^{\binom{k}{2}-1}}{\binom{\binom{k}{2}}{b}}\sum_r\binom{k}{r}\binom{n-k}{k-r}\mathbf{P}(I_iI_\mathbf{1}=1);$$

the 'i' in the last summation above represents any clique for which $|\mathbf{1}\cap i| = r$; $2 \leq r \leq k-1$. Now $I_iI_\mathbf{1}=1$ iff $\mathbf{1}$ and i each have b edges of one color, *which may or may not be the same.* In other words, if we denote by ζ the number of edges in $\mathbf{1}\cap j$, of the color that causes $I_\mathbf{1}$ to equal one, then

$$\mathbf{P}(I_iI_\mathbf{1}=1) = \sum_\zeta\left[\frac{\binom{\binom{k}{2}-\binom{r}{2}}{b-\zeta}\binom{\binom{r}{2}}{\zeta}}{2^{\binom{k}{2}-1}}\right]\frac{1}{2^{\binom{k}{2}-\binom{r}{2}}}\left[\binom{\binom{k}{2}-\binom{r}{2}}{b-\zeta}+\binom{\binom{k}{2}-\binom{r}{2}}{b-\binom{r}{2}+\zeta}\right]$$

$$\leq 2\binom{\binom{k}{2}}{b}^2\frac{1}{2^{k(k-1)-\binom{r}{2}-1}},$$

(2.33)
so that (2.32) yields

$$d_{TV}(\mathcal{L}(X_b),\mathrm{Po}(\lambda_b))$$

$$(2.34)\quad \leq \left(\frac{k^2e}{2b}\right)^b\frac{1}{2^{\binom{k}{2}-1}}\frac{k^2}{2}\left[\frac{ne}{k-2}\right]^{k-2}+2\binom{\binom{k}{2}}{b}\sum_r\frac{\binom{k}{r}\binom{n-k}{k-r}}{2^{\binom{k}{2}-\binom{r}{2}}}$$

$$\leq \left(\frac{k^2e}{2b}\right)^b\frac{1}{2^{\binom{k}{2}-1}}\frac{k^2}{2}\left[\frac{ne}{k-2}\right]^{k-2}+2\binom{\binom{k}{2}}{b}\left[\frac{k^2n}{2^{k-1}}+\frac{k^3\left[\frac{ne}{k-2}\right]^{k-2}}{2^{\binom{k}{2}-1}}\right],$$

as in the proof of Theorem 1. It may now be checked, as before, that the right hand side of (2.34) tends to zero $[k \to \infty, n \leq \frac{\sqrt{2}}{e}k2^{k/2}(1+o(1))]$ only if $b = o(k/\log k)$. This completes the proof of the proposition. $\square$

REFERENCES

ALON, N. AND SPENCER, J. (1992). *The Probabilistic Method,* John Wiley and Sons, Inc., New York.

ARRATIA, R., GOLDSTEIN, L. AND GORDON, L. (1989). *Two moments suffice for Poisson approximations: The Chen-Stein method.* Ann. Probab. **17** 9–25.

ARRATIA, R. AND TAVARÉ, S. (1992). *The cycle structure of random permutations.* Ann. Probab. **20** 1567–1591.

BARBOUR, A., GODBOLE, A. AND QIAN, J. (1993). *On random tournaments.* Under preparation.

BARBOUR, A., HOLST, L. AND JANSON, S. (1992). *Poisson Approximation,* Oxford University Press.

BECK, J. (1991). *An algorithmic approach to the Lovász local lemma I.* Random Structures and Algorithms **2** 343–365.

GRAHAM, R., ROTHSCHILD, B. AND SPENCER, J. (1990). *Ramsey Theory*, 2nd ed., John Wiley and Sons, Inc., New York.

HILDEBRAND, M. (1993). *A Ramsey theory heuristic.* Preprint.

RANDOM WALKS AND UNDIRECTED GRAPH CONNECTIVITY: A SURVEY

ANNA R. KARLIN* AND PRABHAKAR RAGHAVAN†

Abstract. We survey a number of algorithms that decide connectivity in undirected graphs. Our focus is on the use of random walks as a tool in reducing the space complexity of these algorithms.

The *undirected s–t* connectivity (USTCON) problem is the following: given an undirected graph G and two vertices s and t in G, decide whether s and t are in the same connected component. Throughout this paper, n will denote the number of vertices in the graph, and m will denote the number of edges. The USTCON problem is important in the study of space-bounded complexity classes.

In examining algorithms for solving this problem, there are two resource measures of interest.

1. *Time* is measured by the number of steps taken by the algorithm on a standard model of computation such as the unit-cost RAM [1].
2. *Space* is the number of bits of storage used by the algorithm. Here we do not include the read-only storage in which the input to the problem is given.

A classical algorithm for USTCON is *breadth-first search*. Initially, every node of the graph except s is unmarked. Next, we mark s and place it into a queue. Thereafter, the algorithm proceeds by removing the node at the head of the queue and exploring its neighbors one at a time. Each neighbor that is unmarked is now marked, and added to the queue. The algorithm terminates with output YES when t is marked, or with output NO when the queue is empty and t has not been marked. Since each edge is examined at most once in each direction, the time used by the algorithm is clearly $O(m)$. Space is used for marking nodes, and is thus $O(n)$.

Aleliunas, Karp, Lipton, Lovász and Rackoff [3] proposed an alternative algorithm for USTCON using random walks. At each step, the algorithm proceeds from a vertex of G to a neighbor of the vertex chosen uniformly at random. The walk starts at s, and lasts $4m(n-1)$ steps. If t is visited in the course of the walk, the algorithm outputs YES; otherwise, it outputs

* Department of Computer Science and Engineering, FR–35, University of Washington, Seattle, WA 98195. karlin@cs.washington.edu
† IBM T.J. Watson Research Center, Yorktown Heights, NY 10598. pragh@watson.ibm.com

NO. Note that when this algorithm outputs NO, it may do so erroneously. We proceed to bound the probability of such an error.

Let $C_u(G)$ denote the expected length of a walk that starts at u and ends upon visiting every vertex in G at least once. Let $C(G)$ be the *cover time* of G, defined by $C(G) = \max_u C_u(G)$. Aleliunas *et al.* [3] proved the following theorem (see also Göbel and Jagers [11]):

THEOREM 1.

$$C(G) \le 2m(n-1).$$

Proof. Let T be any spanning tree of G. There is a traversal of T, going through vertices $v_0, v_1, \ldots, v_{2n-2} = v_0$ that traverses each edge of T exactly once in each direction. Further, every vertex of G appears at least once in the sequence $v_0, v_1, \ldots, v_{2n-2}$. Consider a random walk that starts at v_0 and terminates upon returning to v_0, having visited the vertices $v_1, v_2, \ldots$ in that order. Since this walk has visited every vertex in G, an upper bound on the expected length of this walk is an upper bound on $C_{v_0}(G)$. Let H_{xy} be the number of steps until state y is first visited starting at state x. Then

$$C_{v_0}(G) \le \sum_{j=0}^{2n-3} H_{v_j, v_{j+1}} = \sum_{\text{edges } (u,w) \in T} H_{uw} + H_{wu}.$$

It is known [2,3,8,11] that

$$H_{vw} + H_{wv} \le 2m, \quad (vw) \in T.$$

But this upper bound holds no matter which vertex of T (and therefore G) we designated to be v_0 in the traversal we chose; therefore $C(G) \le 2m(n-1)$. □

By Theorem 1 and the Markov inequality, we have:

COROLLARY 2. *The probability that the random walk algorithm errs when outputting NO is at most 1/2.*

The random walk algorithm uses space to remember its position in the graph, and to count up to $4m(n-1)$; clearly $O(\log n)$ bits suffice for both of these tasks.

Before proceeding to study the space-time products of the algorithms we have discussed, we pause to observe a feature of the proof of Theorem 1 that will be of interest later on. Chandra *et al.* [8] showed that for any pair of vertices v, w in G,

$$(1) \qquad\qquad H_{vw} + H_{wv} = 2mR_{vw},$$

where R_{vw} is the *effective resistance* between v and w in an electrical network obtained by replacing each edge of G by a unit resistor. Define the distance between v and w to be the effective resistance between them. Let $R_{span}(G)$ denote the weight of the minimum spanning tree in G according to this distance metric. Following the above proof of Theorem 1 and (1), it is clear that $C(G) \leq 2mR_{span}(G)$. We now return to the space-time product of our algorithms.

Interestingly, both the breadth-first search algorithm and the random walk algorithm have the same space-time product, $O(mn)$, within a logarithmic factor. Aleliunas *et al.* [3] posed the following question: "The reachability problem for undirected graphs can be solved in logspace and $O(mn)$ time by a probabilistic algorithm that simulates a random walk, or in linear time and space by a conventional deterministic graph traversal algorithm. Is there a spectrum of time-space trade-offs between these extremes?" The remainder of this paper will be devoted to this question.

For simplicity of presentation, we introduce the following notation. We say that $f(n) = \tilde{O}(g(n))$ if there exist positive numbers c and N such that, for all $n \geq N$, $|f(n)| \leq \log^c n |g(n)|$. Thus, the space-time product achieved by both the above algorithms is $\tilde{O}(mn)$.

Broder, Karlin, Raghavan and Upfal [6] took the first step towards answering the question posed by Aleliunas *et al.* Their algorithm is based on the following observation: the fraction of "wasted" steps of the random walk (steps in which the walk revisits previously visited vertices) increases as the walk grows longer. Hence many short random walks are more efficient in exploring a graph than one long random walk.

The scheme proposed by Broder *et al.* is the basis for the algorithms we discuss below. We now outline their scheme, using space $\tilde{O}(p)$. The algorithm consists of repeating the following steps $O(\log n)$ times.

1. Choose p vertices of G at "random" and place a marker on each. Call these vertices, together with s and t, *leaders*.
2. Repeat $O(\log n)$ times: Take a random walk of length τ from each leader. If such a walk connects two leaders, then mark them and all the other leaders known to be connected to them, as belonging to the same component. If at any point s and t are marked as being in the same component, then stop and output YES.

If in the course of the algorithm, s and t are not marked as being in the same component, output NO. Clearly the space used by the algorithm is $\tilde{O}(p)$. The time requirement is $\tilde{O}(p\tau)$.

The random choice in Step 1 and the value of τ are what distinguish the three algorithms we discuss:

- Algorithm 1: The p vertices are chosen according to the stationary distribution of the random walk, and τ is set to be $\tilde{O}(m^2/p^2)$.
- Algorithm 2: $p/2$ vertices are chosen according to the stationary distribution, and $p/2$ are chosen uniformly at random. The pa-

rameter τ is set to $\tilde{O}(m^{3/2}n^{1/2}/p)$.

- Algorithm 3: $p/2$ vertices are chosen according to the stationary distribution, and $p/2$ are chosen in inverse proportion to vertex degrees. The parameter τ is set to $\tilde{O}((m \sum_v \text{degree}(v))/p)$.

There are two key facts that must be proven in order to show that these algorithms work. The first is to show that a set of p random walks of length τ, one from each of the randomly chosen leaders, visits all the vertices of a connected graph with high probability. Otherwise an adversary could choose s and t among those vertices unlikely to be visited from the other leaders and conceivably foil the algorithm. In other words, we need to derive a bound on the expected time required by p parallel and independent random walks to cover the graph, a problem of interest in its own right. This involves proving something about short-term behavior of the Markov chain and coverage of local neighborhoods in a graph.

The second fact to prove is that if s and t are in the same component and enough leaders are chosen within that component, then with high probability s and t are linked up after a small number of walks from each leader. Coverage of the graph as described above does not assure linkage, since s and t may be visited only by walks from two disjoint sets of leaders that are never linked. Furthermore all the vertices in G could be visited by the walks even with s and t in different components.

Broder *et al.* (algorithm 1) choose a vertex of G to be a leader in proportion to its degree. It follows that each of the random walks of length τ begins in the stationary distribution. They also set $\tau = \tilde{O}(m^2/p^2)$. From this they obtain a space-time product of $\tilde{O}(m^2)$.

To aid the intuition of the reader, let us consider the case when G is a simple path on n vertices. For p leaders chosen at random, with high probability, the maximum gap between two leaders is no more than $n \ln n/p$; the expected time to cover this maximum gap is $\Theta(n^2 \log^2 n/p^2)$. Hence $O(\log n)$ trials (random walks of length $O(n^2 \log^2 n/p^2)$ from each leader) will almost surely cover all the gaps between them for a total of $\Theta(n^2 \log^3 n/p)$ steps. Thus each leader "discovers" its closest neighbor leader in both directions, and therefore all leaders are marked as being in the same component.

For the analysis below, we need to look at the random walk in two ways: first, as a Markov chain $X(t)$ where each state is a vertex in G (the vertex process); second, as a Markov chain $Y(t)$ where each state is a directed edge (the edge process). The transition rule for the vertex process is that if $X(t) = v$, then $X(t+1)$ is equally likely to be any of the neighbors of vertex v. The edge process is defined by $Y(t) = [X(t-1), X(t)], t \geq 1$. The stationary distribution of the vertex process, denoted π, is given by $\pi_v = d_v/(2m)$ where d_v is the degree of the vertex v, and the stationary distribution of the edge process, denoted π', is given by $\pi'_{[v,w]} = 1/(2m)$.

In order to prove the above key facts, we show that the following stronger properties hold with high probability.

1. The random walks from the leaders (each of length τ) will together visit every edge of the graph.
2. A random walk of length τ from any edge will hit some leader.

Property 1 clearly implies the first key fact above. From Properties 1 and 2 together, it is possible to prove the second key fact. Intuitively, the algorithm is likely to discover that two leaders a and b are in the same connected component because there is an edge e such that a random walk of length τ from a is likely to hit e, and a random walk of length τ from e is likely to hit b.

The probability that an edge e is visited in a single random walk of length τ is bounded from below as follows.

$$E_\pi[\text{Number of visits to } e \text{ in } \tau \text{ steps}] = \tau/m$$

$$\leq Pr[e \text{ visited}] \cdot E_e[N_e(\tau)],$$

where $E_e[N_e(t)]$ denotes the expected number of times that a random walk of length t starting at e returns to e. Thus,

$$Pr[e \text{ visited}] \geq \frac{\tau/m}{E_e[N_e(\tau)]},$$

and the problem boils down to upper bounding $E_e[N_e(\tau)]$. Using standard results in renewal theory [2] and an elegant result of Carne [7], Broder *et al.* proved that

$$E_e[N_e(\tau)] \leq \tau/m + \tilde{O}(\sqrt{\tau}).$$

Setting τ to a value that is $\tilde{O}(m^2/p^2)$ yields a space-time product that is $\tilde{O}(m^2)$ for all values of p of interest. The random choices of the leaders and the value of τ chosen by Broder *et al.* cannot hope to achieve a space-time product that is $\tilde{O}(mn)$, as the following example shows. Consider the "barbell" graph consisting of two cliques of size $n/3$ each, joined by a path of length $n/3$. Suppose that $p = n^{3/4}$, and that s and t are in different cliques. In order to achieve the space-time product $\tilde{O}(mn)$, τ must be $\tilde{O}(mn/p^2) = \tilde{O}(n^{3/2})$. A random walk of length $\tilde{O}(n^{3/2})$ from one clique is very unlikely to reach the other clique. On the other hand, no leaders are likely to be selected from the vertices on the path.

The above example highlights the need to favor vertices of low degree when selecting leaders. Barnes and Feige [4] (algorithm 2) took the first step in this direction by selecting $p/2$ leaders according to the stationary distribution (as in [6]), and the remaining $p/2$ uniformly at random. They were able to achieve a space-time product that is $\tilde{O}(m^{3/2}n^{1/2})$.

Feige [10] (algorithm 3) achieved the space-time product that is $\tilde{O}(m\hat{R})$, where $\hat{R} = \sum_v \text{degree}(v)$. Since $\hat{R} \leq n/d_{min}$, where d_{min} is the minimum vertex degree in G, this space-time product is $\tilde{O}(mn/d_{min})$. In his algorithm $p/2$ leaders are chosen according to the stationary distribution, and $p/2$ are chosen in inverse proportion to vertex degrees. The key step in Feige's analysis is to show that

$$(2) \qquad E_e[N_e(\tau)] = \tilde{O}(\sqrt{\tau\hat{R}/m}).$$

There are three crucial points to proving (2). The first is an observation made in [4], that Feige's leader-distribution scheme on G can be analyzed by considering the Broder *et al.* scheme on a modified graph H. The difference between H and G is that self loops are introduced in H.

The second point is to use a method developed by Aldous [2], in his proof that on regular graphs, $E_v[N_v(t)] = O(\sqrt{t})$. Aldous' technique can be applied to the analysis of $E_e[N_e(t)]$ on arbitrary graphs, and can take into account parameters such as d_{min}.

The third point is that $\hat{R}$ is a property of a graph as a whole, whereas we are analyzing properties of short random walks that only visit a small fraction of the graph. Feige's pebble distribution scheme and careful analysis allows him to show that $\hat{R}$ is in fact reflected in the properties of short random walks.

Is there scope for improving Feige's space-time product of $\tilde{O}(m\hat{R})$? Of related interest is the question: how well does the quantity $m\hat{R}$ approximate $\mathcal{C}(G)$? Coppersmith, Feige and Shearer [9] have shown that $R_{span}(G)$ and $\hat{R}$ are within constant factors of each other, for every connected graph. Thus, Feige achieves a space-time product that matches that achieved by the Aleliunas *et al. upper bound* for the random walk. However, there are graphs for which this upper bound is weak. On these, Feige's choice of τ results in a space-time product that is weaker than that of the random walk algorithm.

On the other hand, Feige's algorithm can do better than breadth-first search. For example, on graphs that are regular of degree cn, breadth-first search achieves a space-time product that is $\Omega(n^3)$. In contrast, Feige's algorithm achieves a space-time product that is $\tilde{O}(n^2)$ on such a graph.

What are the intrinsic limits to algorithms that choose leaders at random and then take short random walks from each leader? Feige [10] shows that a class of leader distribution schemes including all those described here cannot achieve a space-time product that is $o(mD)$, where D is the sum of the diameters of the connected components of G. For a survey of other negative results, the reader is referred to Borodin [5]. Wigderson [12] gives a comprehensive survey of additional related results.

REFERENCES

[1] A.V. Aho, J.E. Hopcroft, and J.D. Ullman. *The Design and Analysis of Computer Algorithms.* Addison-Wesley, 1974.

[2] D.J. Aldous. Reversible Markov chains and random walks on graphs, 1993. Unpublished Monograph, Berkeley.

[3] R. Aleliunas, R.M. Karp, R.J. Lipton, L. Lovász, and C. Rackoff. Random walks, universal traversal sequences, and the complexity of maze problems. In *Proceedings of the 20th Annual Symposium on Foundations of Computer Science,* pages 218–223, San Juan, Puerto Rico, October 1979.

[4] G. Barnes and U. Feige. Short random walks. In *Proceedings of the 25th Annual ACM Symposium on Theory of Computing,* pages 728–737, 1993.

[5] A. Borodin. Time-space tradeoffs. In *Proceedings of the 4th Annual ISAAC,* pages 209–220. Springer-Verlag, 1993.

[6] A.Z. Broder, A.R. Karlin, P. Raghavan, and E. Upfal. Trading space for time in undirected s-t connectivity. In *Proceedings of the 21st Annual ACM Symposium on Theory of Computing,* pages 543–549, Seattle, WA, May 1989. ACM.

[7] T.K. Carne. A transmutation formula for Markov chains. *Bull. Sci. Math.,* 109:399–405, 1985.

[8] A.K. Chandra, P. Raghavan, W.L. Ruzzo, R. Smolensky, and P. Tiwari. The electrical resistance of a graph captures its commute and cover times. In *Proceedings of the 21st Annual ACM Symposium on Theory of Computing,* pages 574–586, Seattle, May 1989.

[9] D. Coppersmith, U. Feige, and J. Shearer. Random walks on regular and irregular graphs. Technical Report CS93-15, The Weizmann Institute of Science, 1993.

[10] U. Feige. A randomized time-space tradeoff of $\tilde{O}(m\hat{R})$ for USTCON. In *34th Annual IEEE Symposium on Foundations of Computer Science,* pages 238–247, 1994.

[11] F. Göbel and A.A. Jagers. Random walks on graphs. *Stochastic Processes and their Applications,* 2:311–336, 1974.

[12] A. Wigderson. The complexity of graph connectivity. Unpublished manuscript, Computer Science Dept., Hebrew University, 1993.

SIDON SETS WITH SMALL GAPS

JOEL SPENCER* AND PRASAD TETALI[†]

1. Introduction. Let $\mathcal{N}$ denote the set of positive integers. A set $\mathcal{A} \subset \mathcal{N}$ is called a *Sidon set* if the sums $a + a'$ $(a, a' \in \mathcal{A})$ are all distinct. For background on Sidon sets, we refer to [6]. For additional recent results see [2], [3]. In this paper we are interested in infinite Sidon sets, thus for the rest of the discussion all Sidon sets are infinite unless, otherwise specified. We denote the *sumset* $\mathcal{A} + \mathcal{A}$ by S_A. i.e. for a Sidon set $\mathcal{A}$,

$$S_A = \{s_1, s_2, \ldots\} = \{a + a' : a, a' \in \mathcal{A}\}.$$

Erdős et al. [3] consider the problem of estimating the size of the gaps between consecutive elements of S_A. In particular they prove the following result.

THEOREM 1.1 (ERDŐS ET AL.). *There is an infinite Sidon set $\mathcal{A}$ such that the sumset $S_A = \{s_1, s_2, \ldots\}$ satisfies, with $0 < \epsilon < 1$,*

$$s_{i+1} - s_i < s_i^{1/2}(\log s_i)^{3/2+\epsilon} \quad (\text{for } i > i_0)$$

In this paper we improve the above result as follows.

THEOREM 1.2. *There exists an (infinite) Sidon set such that*

$$s_{i+1} - s_i < cs_i^{1/3} \ln s_i \quad (\text{for } i = 1, 2, \ldots).$$

where $c > 0$ is an absolute constant.

The main idea is probabilistic and is quite simple. The proof makes crucial use of Janson's (correlation) inequality. However, the details get a little technical. The proof techniques are quite similar to those used in [4] (see also Ch. 3 of [5]). Here is the proof idea.

Suppose we want to show the existence of a Sidon set whose sumset always has an element in the interval I from n to $n + cn^{1/3} \ln n$ with c large (say, $c = 6 \times 10^{10}$). Let S be a random set with $\Pr[x \in S] = c_1 x^{-2/3}$ and c_1 small (say, $c_1 = .01$). Whenever $x + y = a + b$ with $x, y, a, b \in S$ remove the *maximal* of x, y from S, let S^- be the set of remaining elements. So S^- is tautologically Sidon and, by Borel-Cantelli, one "only" needs that the probability of there being no $x, y \in S^-$ with $x + y \in I$ is, say, $O(n^{-2})$. Using Janson's Inequality we get that w.h.p. (with high probability) there

---
* spencer@cs.nyu.edu, Courant Institute of Mathematical Sciences, 251 Mercer St., New York, NY 10012.

† prasad@research.att.com, 600 Mountain Ave., Room 2C-169, AT & T Bell Labs, Murray Hill, NJ 07974.

are $x, y \in S$ with $x + y \in I$ and (with more work) that the number of *distinct* such pairs is w.h.p. "nearly" the expectation. Now the expected number of x, y, a, b, c with all in S, x, y as above and $x + a = b + c, a < x$ is (with c_1 small) a small fraction of the previous expectation and one needs that w.h.p. the number of such 5-tuples with *distinct* x, y is relatively close to its expectation. This last requirement while certainly true, demands bit of a (technical) calculation.

2. The proof. Consider a random set S as follows. For each $x \in \mathcal{S}$,

$$\Pr[x \in S] = c_1 x^{-2/3}, \quad c_1 = .01 \text{ (say)}.$$

Let $I = [n, n + cn^{1/3} \ln n]$, with $c = 6 \times 10^{10}$. We call (x, y) a *pair* if $x, y \in S$, $\frac{n}{3} \leq x < y \leq \frac{2n}{3}$, and $x + y \in I$. Then

$$\mu = E[\# \text{ pairs }] = \frac{n}{3} \cdot cn^{1/3} \ln n \cdot (c_1 n^{-2/3})^2 = \frac{cc_1^2}{3} \ln n,$$

since there are $n/3$ choices for x, and $cn^{1/3} \ln n$ choices for y, and each is picked with probability $c_1 n^{-2/3}$, independently.

First act—distinct pairs

We call pairs (x, y) and (x', y') *distinct* if $\{x, y\} \cap \{x', y'\} = \phi$. We want to prove that the number of distinct pairs is close to μ. Towards this let us first estimate the correlation term, $\Delta = \sum \Pr[(x, y) \text{ and } (x', y') \text{ exist}]$, where the summation is over *nondistinct* pairs.

$$\Delta = O\left(n \cdot (cn^{1/3} \ln n)^2 \cdot (c_1 n^{-2/3})^3\right)$$

(since we get to choose only 3 out of x, y, x', y')

$$= O(n^{-1/3 + o(1)})$$

We are now in the framework of Lemma 4.2 of [1], which is a consequence of Janson's Inequality. Following the notation of [1], by a *maxdisfam*, we mean a maximal collection (or family) of distinct pairs. Then by Lemma 4.2 and due to the fact that $\Delta = o(1)$, we have

$$\Pr[\text{there exists a maxdisfam } J, |J| = s] \leq \frac{\mu^s}{s!} e^{-\mu + o(1)}$$

Then for a fixed $0 < \epsilon < 1$, we have

$$\Pr[\text{there exists a maxdisfam } J, |J| \leq (1 - \epsilon)\mu]$$

$$\leq \sum_{s=0}^{(1-\epsilon)\mu} \frac{\mu^s}{s!} e^{-\mu + o(1)}$$

$$= e^{-\mu + o(1)} \sum_{s=0}^{(1-\epsilon)\mu} \frac{\mu^s}{s!}$$

$$< e^{-\mu + o(1)} \left(\frac{e}{(1-\epsilon)}\right)^{(1-\epsilon)\mu}, \quad \text{(note } 0 < \epsilon < 1\text{)}.$$

Clearly, we can choose c and c_1 (in fact, we did) such that $e^{-\mu+o(1)} < e^{(-3+o(1))\ln n}$. Also, since $(\frac{e}{(1-\epsilon)})^{1-\epsilon} \to 1$, as $(1-\epsilon) \to 0^+$, we can fix ϵ such that $\left(\frac{e}{(1-\epsilon)}\right)^{(1-\epsilon)\mu} < e^{\ln n}$. Thus

$$Pr[\text{there exists a maxdisfam } J, |J| \leq (1-\epsilon)\mu] \leq e^{(-3+o(1))\ln n} e^{\ln n}$$
$$= n^{-2+o(1)}.$$

This allows us to conclude, by Borel-Cantelli lemma, there exists a fixed $0 < \epsilon < 1$, such that

$$\text{a.a. the number of distinct pairs in } S > (1-\epsilon)\mu \qquad (*)$$

Second act—deletion of overlaps

We construct a Sidon set out of S as follows. Whenever $x + y = a + b$ with $x, y, a, b \in S$, remove the *maximal* of $\{x, y\}$ from S. Let A be the set of remaining elements. By our construction, A is a Sidon set. The rest of the paper is devoted to proving the assertion (in Theorem 1.2) on the gaps in the sumset S_A.

We call a pair (x, y) *bad* if there exists a 5-tuple (x, y, a, b, c) with $x, y, a, b, c \in S$ where, as before (and for the rest of the paper), we have $\frac{n}{3} \leq x < y \leq \frac{2n}{3}$, $x + y \in I$, and we also have $a < x$, $a + x = b + c$. We are interested in these 5-tuples, since each such tuple makes $x \in S$, but $x \notin A$, thus affecting the number of (*good*) pairs (x, y) with $x, y \in A$, $x + y \in I$. Recall that our main aim is to show that the number of such *good* pairs is positive. In view of (*) above (from First act), it suffices to show that the number of *distinct bad* pairs is strictly smaller than the number of distinct pairs.

We first estimate the expected number of such bad pairs BP.

$$\mu^{bad} = E[BP] = \sum_{\substack{x,y \in I \\ a < x \\ a+x=b+c}} Pr[x, y, a, b, c, \in S]$$

The "crude" estimate is sufficient for the lower bound. Note that there are $n/3$ choices for x, and at least $n/3$ choices for each of a and b. Thus

$$\mu^{bad} \geq (c_1 n^{-2/3})^5 \sum_{\substack{x,y \in I \\ a < x \\ a+x=b+c}} 1$$

$$\geq \left(\frac{n}{3}\right)^3 \cdot cn^{1/3} \ln n \cdot (c_1 n^{-2/3})^5$$

$$\geq (\frac{cc_1^5}{27} \ln n).$$

On other hand, care must be taken to give an upper bound, since $z^{-2/3}$ goes to infinity, as z goes to zero. But the rough calculation is as follows –

there are at most $2n/3$ and $4n/3$ choices for a and b, respectively (having chosen x); therefore,

$$
\mu^{bad} \leq \left(\frac{n}{3} \cdot \frac{2n}{3} \cdot \frac{4n}{3} \cdot cn^{1/3} \ln n \cdot (c_1 n^{-2/3})^5 \right)
$$

$$
\leq \left(\frac{8cc_1^5}{27} \ln n \right).
$$

This would be rigorous if a, b, and c were all "big", e.g. bigger than ϵn. To make this rigorous, we bound the worrysome case as follows: suppose $a < \epsilon n$, and $b < \epsilon' n$. (Note that since x is big, at least one of b and c has to be big; w.l.o.g., we assume c is big.)

$$
\mu^{bad} = \sum_{\substack{x,y \in I \\ a < \epsilon n, b < \epsilon' n \\ a+x=b+c}} \Pr[x, y, a, b, c \in S]
$$

$$
\leq \frac{cc_1^2}{3} \ln n \sum_{\substack{a < \epsilon n, b < \epsilon' n \\ a+x=b+c}} \Pr[a, b, c \in S]
$$

$$
\leq \frac{cc_1^2}{3} \ln n (c_1 n^{-2/3}) \left(\int_{a=0}^{\epsilon n} c_1 a^{-2/3} + O(1) \right) \left(\int_{b=0}^{\epsilon' n} c_1 b^{-2/3} + O(1) \right)
$$

$$
\leq [(\epsilon \epsilon')^{1/3} cc_1^5 + o(1)] \ln n.
$$

The point here is that $\mu^{bad} \ll \frac{cc_1^2}{3} \ln n = \mu$.

Note that this makes $\mu' = \mu - \mu^{bad}$ "close" to μ. So potentially we could remove each such (problematic) x, and still have $c'' \ln n$ good pairs (x, y), with $c'' \approx \frac{cc_1^2}{3}$. For this intuition to prove correct, we need to count the actual number BP and not just $E[BP]$. The difficulty here arises in view of the *correlation* between bad pairs. Distinct bad pairs (x, y) and (x', y') might be correlated due to the overlap of the corresponding 5-tuples. We first show that any maximal *disjoint* collection of such 5-tuples is of size close to μ^{bad}.

LEMMA 2.1. $\Pr[\exists a \text{ disjoint collection of 5-tuples of size } > 10\mu^{bad}] < n^{-2+o(1)}$.

Proof. Straight forward using Lemma 4.1 of [1]. By Lemma 4.1 of [1] we have,

$\Pr[\exists a \text{ disjoint collection of 5-tuples of size } > 10\mu^{bad}]$

$$
< \sum_{s=10\mu^{bad}}^{\infty} \frac{(\mu^{bad})^s}{s!}
$$

$$
< \left[\frac{e}{10} \right]^{10\mu^{bad}} \quad \text{(note that our choice of } c
$$

$$
\text{and } c_1 \text{ makes } 10\mu^{bad} > 2\ln n)
$$

$$
< n^{-2+o(1)}. \qquad \square
$$

Consider the graph with distinct bad pairs as the vertices, and placing an edge between distinct bad pairs (x, y) and (x', y') iff there exist two 5-tuples, (x, y, a, b, c) and (x', y', a', b', c') such that $\{x, y, a, b, c\} \cap \{x', y', a', b', c'\} \neq \phi$. We call such a 5-tuple an *overlapping* 5-tuple, or $O5T$ for short.

Note that Lemma 2.1 guarantees that a.s. the size of a maximal independent set in G is at most $10\mu^{bad}$. Thus, in order to prove that the actual number of distinct bad pairs (which is also the number of vertices in G) is close to its expectation (i.e. μ^{bad}), it suffices to show that the maximum degree of each vertex in G is bounded. The following claim asserts that the expected number of edges, and hence the expected maximum degree is $o(1)$.

Claim. $\mu_{O5T} = E[\#O5T's] = n^{-1/3+o(1)}$.

We defer the proof of this claim to the end of this section. For now we continue with the proof of Theorem 1.2, assuming the claim. The following lemma bounds the size of a maximal disjoint collection of $O5T$'s. (The only overlaps in such a collection are between the two 5-tuples that form each $O5T$.)

LEMMA 2.2. *a.a. there exists an (absolute constant) L such that the number of disjoint $O5T's \leq L$.*

Proof. By now the proof technique should be familiar.

$$\Pr[\# \text{ of disjoint } O5T's > L] < \frac{(\mu_{O5T})^L}{L!}$$
$$< \frac{1}{L!}\left(n^{-1/3+o(1)}\right)^L$$
$$< n^{-2+o(1)}, \quad \text{for } L \geq 6.$$

Once again, by Borel-Cantelli, we have the lemma.   □

Note that with the above proof technique, we can also prove the following. Consider an arbitrary subset H of a possible 5-tuple such as (x, y, a, b, c). Then we can prove that the size of a maximal collection of 5-tuples, with H as the pairwise intersection of the 5-tuples, is also a.a. bounded by an absolute constant L'. One can now use the technique from [4], which uses the Erdős-Rado Δ-system lemma, to show that the maximum degree in G is bounded by L'', yet another constant dependent only on L'.

Proof of Theorem 1.2. Lemmas 2.1 and 2.2 together with the above arguments imply that

a.a. for every large n, the number of distinct bad pairs $(BP's)$

$$< 10\mu^{bad} + C,$$

where C is an absolute constant. However, the number of distinct pairs, by (*), is at least $(1 - \epsilon)\mu > 10\mu^{bad} + C$. This implies that in going from S to the Sidon set A, we have not lost too many "good" pairs. Thus A is Sidon, and has the property that for every large n, the sumset of A always has an element in $[n, n + cn^{1/3} \ln n]$, where $c = 6 \times 10^{10}$. $\qquad\square$

Proof of claim. We want to estimate the expected number of overlapping 5-tuples $(O5T's)$. Let T and T' represent $\{x, y, a, b, c\}$ and $\{x', y', a', b', c'\}$ respectively. By definition,

$$\mu_{O5T} = \sum_l \sum_{|T \cap T'|=l} \Pr[T, T' \text{ is an } O5T]$$

where l varies over the possible values for the size of the intersection of the two 5-tuples, given that (x, y) and (x', y') are distinct pairs. It turns out (and we leave it to the avid reader to check all the cases) that the important cases to check are when $l = 1$, the overlap being the least. We consider the following cases which we hope exemplify the kind of calculation that is to be checked. More importantly, we hope these cases serve as an outline of the proof of the above claim.

The first thing to notice is that once we choose 3 out of $\{a, x, b, c\}$, the 4th element is determined due to $a + x = b + c$.

Case 2.3. One of $\{a, b, c, a', b', c'\}$ equals one of $\{x, x', a, a', b, b', c, c'\}$, in a nontrivial way. Without loss of generality, let $a = b'$. Then the contribution to $\mu_{O5T} = O(n^5(cn^{1/3} \ln n)^2(c_1 n^{-2/3})^9) = O(n^{-1/3+o(1)})$. Here we used the fact that we have O(n) choices for all the elements except for y and y', and that the overlap is exactly one element.

Case 2.4. One of $\{a, b, c, a', b', c'\}$ equals one of $\{y, y'\}$. Let $a = y'$. In this case, we count the choices for y' first, having chosen x and then we count the rest. Thus the contribution is, once again, $O(n^5(cn^{1/3} \ln n)^2(c_1 n^{-2/3})^9) = O(n^{-1/3+o(1)})$.

The remaining cases follow similarly. $\qquad\square$

Remark 2.5. The constant c in our main theorem is quite large. At the same time, we had a lot of room to spare in the analysis of the proof; since we believe the real problem is in lowering the $n^{1/3} \ln n$ – bound, we make no attempt at tightening the constants.

REFERENCES

[1] N. Alon, P. Erdős, and J. Spencer, *The Probabilistic Method*, John Wiley, New York, 1992.

[2] P. Erdős and R. Freud, On sums of a Sidon-sequence, *J. of Number Theory* **38** (1991), 196-205.

[3] P. Erdős, A. Sárközy, and V.T. Sós, On sum sets of Sidon sets, I, Preprint No. 32 (1993), Mathematical Institute of the Hungarian Academy of Sciences.

[4] P. Erdős and P. Tetali, Representations of integers as the sum of k terms, *Random Structures and Algorithms*, **1** (1990), 245-261.

[5] P. Tetali, *Analysis and Applications of Probabilistic Techniques*, Ph.D. Thesis (1991), New York University.

[6] H. Halberstam and K.F. Roth, *Sequences*, Springer-Verlag, Berlin-Heidelberg-New York (1983).

VARIATIONS ON THE MONOTONE SUBSEQUENCE THEME OF ERDŐS AND SZEKERES*

J. MICHAEL STEELE†

Abstract. A review is given of the results on the length of the longest increasing subsequence and related problems. The review covers results on random and pseudo-random sequences as well as deterministic ones. Although most attention is given to previously published research, some new proofs and new results are given. In particular, some new phenomena are demonstrated for the monotonic subsequences of *sections* of sequences. A number of open problems from the literature are also surveyed.

Key words. Monotone subsequence, unimodal subsequence, partial ordering, limit theory, irrational numbers, derandomization, pseudo-random permutations.

AMS(MOS) subject classifications. 60C05, 06A10

1. Introduction. The main purpose of this article is to review a number of developments that spring from the classic theorem of Erdős and Szekeres (1935) which tells us that from a sequence of $n^2 + 1$ distinct real numbers we can always extract a *monotonic* subsequence of length at least $n + 1$. Although the Erdős-Szekeres theorem is purely deterministic, the subsequent work is accompanied by a diverse collection of results that make contact with randomness, pseudo-randomness, and the theory of algorithms.

Central to the stochastic work that evolved from the discovery of Erdős and Szekeres is the theorem that tells us that for independent random variables X_i, $1 \leq i < \infty$, with a continuous distribution, the length I_n of the longest increasing subsequence in $\{X_1, X_2, \ldots, X_n\}$,

$$I_n = \max\{k : X_{i_1} < X_{i_2} < \ldots < X_{i_k} \text{ with } 1 \leq i_1 < i_2 < \ldots < i_k \leq n\},$$

satisfies

$$(1.1) \qquad \lim_{n \to \infty} I_n/\sqrt{n} = 2 \quad \text{with probability one .}$$

The suggestion that 2 might be the right limiting constant was first put forth by Baer and Brock (1968) who had engaged in an interesting Monte Carlo study motivated by S. Ulam (1961), but the first rigorous progress is due to Hammersley (1972) who showed $I_n/\sqrt{n}$ converges in probability to a constant $C > 0$. The constant C proved to be difficult to determine, but Logan and Shepp (1977) first showed $C \geq 2$ by a sustained argument of the calculus of variations and then $C \leq 2$ was established by Vershik and Kerov

* Research supported in part by NSF Grant Number DMS 92-11634 and Army Research Office Grant DAAL03-91-G-0110.

† Department of Statistics, University of Pennsylvania, The Wharton School, 3000 Steinberg Hall-Dietrich Hall, Philadelphia, PA 19104-6302.

(1977) using information about the Plancherel measure on Young-tableaux. The later result was subsequently simplified by Pilpel (1986) who also gave the bound valid for all n,

$$EI_n \leq \sum_{j=1}^{n} 1/\sqrt{j}.$$

More recently, Aldous and Diaconis (1993) have given an insightful proof that $C = 2$ by building on elementary aspects of the theory of interacting particle systems.

In the next section we sketch six (or more) proofs of the Erdős-Szekeres theorem. The central intention of reviewing these proofs is to see what each of the methods can tell us about combinatorical technique, but the very multiplicity of proofs of the Erdős-Szekeres theorem offers an indication that the result is not so special as one might suspect. Quite to the contrary, the Erdős-Szekeres theorem offers us a touchstone for understanding a variety of useful ideas. The third section looks at generalizations of (1.1) to $d \geq 2$ and to structures like unimodality. Section four then develops the theory of monotone subsequences for pseudo-random sequences. That section gives particularly detailed information about the Weyl sequences, $x_n = n\alpha \bmod 1$ for irrational α. Section five develops the theory of monotone subsequences for sections of an infinite sequence.

The final section comments briefly on open problems and on some broader themes that seem to be emerging in the relationship between probability and combinatorial optimization.

2. Six or more proofs. Perhaps the most widely quoted proof of the Erdős-Szekeres theorem is that of Hammersley (1972) which uses a visually compelling pigeon-hole argument. The key idea is to place the elements of the sequence $x_1, x_2, \ldots, x_m$ with $m = n^2 + 1$ into a set of ordered columns by the following rules:

(a) let x_1 start the first column, and, for $i \geq 1$,

(b) if x_i is greater than or equal to the value that is on top of a column, we put x_i on top of the first such column, and

(c) otherwise start a new column with x_i. The first point to notice about this construction is that the elements of any column correspond to an increasing subsequence. The second observation is that the only time we shift to a later column is when we have an item that is smaller than one of its predecessors. Thus, if there are k columns in the final structure, we can trace back from the last of these and find monotone decreasing subsequence of length k. Since $n^2 + 1$ numbers are placed into the column structure, one must either have more than n columns or some column that has height greater than n. Either way, we find a monotone subsequence of length $n + 1$.

Hammersley's proof is charming, but the original proof of Erdős and Szekeres (1935) can in some circumstances teach us more. To follow the original plan, we first let $f(n)$ denote the least integer such that any sequence of $f(n)$ real numbers must contain a monotone subsequence of length n. One clearly has $f(1) = 1$, $f(2) = 2$, and, with a moments reflection, $f(3) = 5$. By using the construction that extends the example $\{3, 2, 1, 6, 5, 4, 9, 8, 7\}$ we see $f(n) > (n-1)^2$, and the natural conjecture is that $f(n) = (n-1)^2 + 1$. The method we use to prove this identity calls on a modest bit of geometry, but the combinatorial technique that it teaches best could be called the *abundance principle*: In many situations if a structure of a certain size must have a special substructure, then a somewhat larger structure must have many of the special substructures.

The Erdős-Szekeres proof is quickly done at a blackboard, although a few more words are needed on paper. To show by induction that $f(n) = (n-1)^2 + 1$, we begin by considering an integer $b \geq 0$ and a set of $f(n) + b$ distinct points in $\mathbb{R}^2$. By applying the induction hypothesis $b + 1$ times we can find $b + 1$ distinct points that are terminal points of monotone sets of length n. For the moment we withhold our budget set B of b points, and we invoke the induction hypothesis on set P with cardinality $f(n)$. The induction hypothesis gives us a monotone subsequence of length n. We remove the last point of this sequence from P, and we add one of the budget points from B to get a new set P' of cardinality $f(n)$.

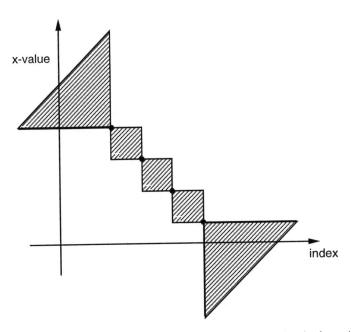

FIG. 2.1. *The dots denote the points of S^+, and the shaded region is the region D of points that are not comparable to the points of S^+ in the up-and-to-the-right order.*

Now, just to get started, suppose we took $b = 2n$, then since any terminal point must be associated with an decreasing sequence or an increasing sequence, we can suppose without loss of generality that there are $n + 1$ points S that are terminal points of increasing sequence of length n. If some two elements of S were in increasing order, we could extend one of the length n increasing sequences to a sequence of length $n + 1$. On the other hand if no two elements of S are in increasing order, then S constitutes a set of size $n + 1$ in decreasing order. We have therefore shown $f(n + 1) \leq f(n) + 2n$, which is enough to show that $f(n) \leq n^2 + 1$, but we can do better by exercising a little more care.

This time we add just $b = 2n - 1$ new points and get $2n$ terminal points. If we have $n + 1$ of either of the two types of terminals, then we can proceed as before to find a monotone sequence of length $n + 1$. Therefore we can suppose that there are exactly n terminal points of each type. If S^+ and S^- denote the terminals of the increasing and decreasing sequences, then by the argument of the previous paragraph there is no loss in assuming that S^+ forms a decreasing set (see Figure 2.1). Now if any point $x \in S^-$ were to be up-and-to-the-right of any point of $y \in S^+$, then x would combine with the increasing sequence sending y to form an increasing sequence of length $n + 1$. Therefore all points of S^- are in the region D. These are an increasing sequence of length n, and the last of there points is majorized by some $y \in S^+$, telling us $S^+ \cup \{y\}$ would give the required subsequence.

More will be said later about the virtue of the Erdős-Szekeres proof, but first we want to recall what is perhaps the slickest and most systematic proof, the one due to Seidenberg (1959) and which is naturally suggested by dynamic programming. We take $S = \{p_1, p_2, \ldots, p_m\} \subset \mathbb{R}^2$ distinct points ordered by their x-coordinates, and we define $\varphi : S \to \mathbb{Z} \times \mathbb{Z}$ by letting $\varphi(p) = (s, t)$ if s is the length of the longest increasing subsequence terminating at p, and t is the length of the longest decreasing subsequence terminating at p. We are not too concerned with algorithms in this review, but at this point one may as well note that there is no problem in calculating $\varphi(p_k)$ in time $0(k)$ given $\varphi(p_1), \varphi(p_2), \ldots, \varphi(p_{k-1})$ so the complete computation of φ on S can be determined in time $0(m^2)$, (cf. Friedman (1975)). Now, if S contains no monotone subsequence of length n, then $\varphi(S) \subseteq \{1, 2, \ldots, n - 1\} \times \{1, 2, \ldots, n - 1\}$. But φ is injective, since, if $p, q \in S$ and q follows p in x-coordinate order, then $\varphi(q)$ might have at least one coordinate larger than the corresponding coordinate of $\varphi(p)$. Hence we see that if $m \geq (n - 1)^2 + 1$ we find that S must contain a monotone subsequence of length n. In other words, we have $f(n) \geq (n - 1)^2 + 1$ and thus complete our third proof of the Erdős-Szekeres Theorem.

The fourth proof we consider is one due to Blackwell (1971) that is not so systematic as that of Seidenberg (1959) or as general the original proof of Erdős and Szekeres, but it serves well to make explicitly the connection to greedy algorithms.

If $S = \{x_1, x_2, \ldots, x_r\}$ is our set of $r > nm$ distinct real numbers, we say a monotone decreasing subsequence S' is *leftmost* if $x'_1 = x_1$ and each term x'_i of S' is equal to the next term of S' which is smaller than x'_{i-1}. Thus S' is the consequence of applying a greedy algorithm to the sequence S.

If we successively apply this greedy process to the points that remain after removal of the leftmost decreasing subsequence we obtain a decomposition of S into $S_1, S_2, \ldots, S_t$ where each S_i is a decreasing subsequence. The observation about this decomposition is that we can construct an increasing subsequence $\{a_1, 2, \ldots, a_t\}$ of S by the following backward moving process:

1. Select a_t arbitrarily from S_t
2. For any $j = t$ down to 1, select a_{j-1} as any term in S_{j-1} that is smaller than a_j.

Because of the definition of the S_j, $1 \leq j \leq t$ we can always complete the steps required in this process. We then find either an increasing set $\{a_1, a_2, \ldots, a_t\}$ with $t > n$ or else one of the decreasing subsequences S_j has cardinality bigger than m. In retrospect one can see that Blackwell's proof is almost isomorphic to Hammersley, though the associated picture and algorithmic feel are rather different.

The fifth proof is closely related to the one just given, but it still offers some useful distinctions. In the solution of Exercise 14.25, Lovász (1979) suggests that given a set S of $n^2 + 1$ real numbers $\{x_1, x_2, \ldots, x_{n^2+1}\}$ one can define a useful partition $A_1, A_2, \ldots$ of S by taking A_k to be the set of all x_j with $1 \leq j \leq n^2 + 1$ for which the longest increasing subsequence beginning with x_j has length exactly equal to k. One can easily check from this definition that each of the sets $A_k = \{i_1 < i_2 < \ldots < i_s\}$ gives rise to k monotone decreasing subsequence $x_{i_1} > x_{i_2} > \ldots > x_{i_s}$, and from this observation the Erdős-Szekeres theorem follows immediately. This last proof has the benefit of showing that any digraph with no directed path of length greater than k has chromatic number bounded by k.

The fifth proof is one that deserves serious consideration, but which would lead us too far afield for us to develop in detail. The central idea is that of the construction of Schensted (1961) that provides a one-to-one correspondence between pairs of standard Young tableaux and the set of permutations. This correspondence as well as its application to the theorem of Erdős and Szekeres and to algorithms for the determination of the longest monotone subsequence of a permutation are well described in Stanton and White (1986). The work of Schensted (1961) was substantially extended by Knuth (1970) to objects that go well beyond permutations. We do not know the extent to which the correspondence provided by Knuth (1970) might contribute to the central problems of this review.

Our final observation on the proofs is just to note that the Erdős-Szekeres theorem also follows from the well known decomposition theorem of Dilworth (1950) which says that any finite partially ordered set can be partitioned into k chains $C_1, C_2, \ldots, C_k$ where k is the maximum cardinality of all anti-chains in S. Using our previous S with the up-and-to-the-right ordering, we see that if there is no decreasing subsequence of length n, then $k < n$ and

$$|C_1| + |C_2| + \ldots + |C_k| = |S| \geq (n-1)^2 + 1$$

implies that for some $|C_j|$ we have $|C_j| \geq n$. Since C_j corresponds to an increasing sequence, we have the final proof.

Dilworth's theorem has itself been the subject of many further developments, some of which are directly connected to the issues engaged by this review. For a recent survey of the work related to Dilworth's theorem and sketches of several proofs of Dilworth's theorem, one should consult Bogart, Greene, and Kung (1990). A result that generalizes both the Dilworth decomposition theorem and the digraph theorem of the previous paragraph is the theorem of Gallai and Milgram (1960) which says that if α is the largest number of points in a digraph G that are not connected by any edge, then G can be covered by α directed paths.

3. Higher dimensions. The charm of the Erdős-Szekeres monotonicity theorem goes beyond the call for innovative proofs. There are several useful extensions and generalizations, though amazingly almost all of these have grown up a good many years after the original stimulus.

One natural issue concerns the generalization of the monotonicity theorem to d-dimensions. If we define a partial order on $\mathbb{R}^d$ by saying $y << w$ if $y = (y_1, y_2, \ldots, y_d)$ and $w = (w_1, w_2, \ldots, w_d)$ satisfy $y_1 \leq w_1$, $y_2 \leq w_2, \ldots, y_d \leq w_d$, then the natural variables of interest are $\Lambda^+(y_1, y_2, \ldots, y_n)$, the length of the longest chain in the set $\{y_1, y_2, \ldots, y_n\} \subset \mathbb{R}^d$ and correspondingly $\Lambda^-(y_1, y_2, \ldots, y_n)$, the cardinality of the largest anti-chain in $\{y_1, y_2, \ldots, y_n\}$. After tentative steps in Steele (1977) which were nevertheless enough to settle a conjecture of Robertson and Wright (1974), the definitive result was established by Bollobás and Winkler (1988). Their main result is that for $\{X_i : 1 \leq i < \infty\}$, independent and uniformly distributed on $[0, 1]^d$, one has

$$\lim_{n \to \infty} \Lambda^+(X_1, X_2, \ldots, X_n)/n^{1/d} = c_d > 0$$

where the limit holds with probability one and where c_d is a constant that depends on the dimension $d \geq 1$. Further, Bollobás and Winkler (1988) also showed that the constants c_d satisfy the interesting relation

$$\lim_{d \to \infty} c_d = \sum_{n=0}^{\infty} \frac{1}{n!} = e.$$

Since it is trivial that $c_1 = 1$ and since we know from Logan and Shepp (1977) and Vershik and Kerov (1977) that $c_2 = 2$, we thus have in hand three instances that support the humorous but feasible speculation due to

Aldous (personal communication) that perhaps

$$c_d = \sum_{n=0}^{d} \frac{1}{n!}$$

for all $d \geq 2$. The correctness of this interpolation and also the one based on $c_d = (1 + 1/(d+1))^{d-1}$ are placed on slippery ground by simulations of P. Winkler and R. Silverstein (cf. Silverstein (1988)) that suggest that c_3 is approximately 2.35, a value that does not agree well with either of the two candidates of 5/2 and 9/4. Still, as the investigators note the convergence in the simulations was very slow, and the value of 2.35 is the result of heuristic curve fitting. The determination of c_d remains of interest from several points of view.

4. Totals and functionals of totals. Lifschitz and Pittel (1981) have studied the total number T_n of increasing subsequences of n independent uniformly distributed random variables. Among other results they found

$$ET_n \sim \alpha n^{-1/4} e^{2\sqrt{n}}$$

where $\alpha = (2\sqrt{\pi e})^{-1}$, and

$$E(T_n^2) \sim \beta_0 n^{-1/4} \exp(\beta_1 n^{1/2})$$

where $\beta_1 = 2(2 + \sqrt{5})^{1/2}$ and β_0 is equal to $(20\pi(2 + \sqrt{5})^{1/2} \exp(2 + \sqrt{5}))^{-1/2} \sim 0.016$. These asymptotic results were obtained by analytic methods beginning with the easy formula

$$E(T_n) = \sum_{k=0}^{n} \frac{1}{k!} \binom{n}{k}$$

and the trickier

$$E(T_n^2) = \sum_{k+\ell \leq n} r^\ell \left\{ \frac{1}{(k+\ell)!} \right\} \binom{n}{k+\ell} \binom{(k+1)/2 + \ell - 1}{\ell}.$$

Lifschitz and Pittel (1981) also proved that there is a constant γ such that as $n \to \infty$

$$n^{-\frac{1}{2}} \log T_n \to \gamma$$

in probability and in mean. The exact determination of γ remains an open problem, though one has the bounds

$$2 \log 2 \leq \gamma \leq 2.$$

5. Unimodal subsequences. The most natural variation on the theme of monotone subsequences is perhaps that of unimodal subsequences, which are those that either increase to a certain point and then decrease, or else decrease to a point and increase thereafter. Formally, given a sequence $S = \{x_1, x_2 \ldots, x_n\}$ we are concerned with $U_n(S)$ defined by $U_n(S) = \max(U_n^+, U_n^-)$ where

$$U_n^+ = \max\{k : \exists 1 \leq i_1 < i_2 < \ldots < i_k \text{ with}$$
$$x_{i_1} < x_{i_2} < \ldots < x_{i_j} > x_{i_{j+1}} > \ldots > x_{i_k} \text{ for some } j\}$$

and

$$U_n^- = \max\{k : \exists 1 \leq i_1 < i_2 < \ldots < i_k \text{ with}$$
$$x_{i_1} > x_{i_2} > \ldots x_{i_j} < x_{i_{j+1}} < \ldots < x_{i_k} \text{ for some } j\}$$

In a remarkable *tour-de-force*, Chung (1980) established the result that for any sequence S of n distinct reals, we have

$$(5.1) \qquad U_n \geq \lceil (3n - 3/4)^{1/2} - 1/2 \rceil,$$

and, moreover, this result is best possible in the sense that for any $n \geq 1$ there is a sequence of distinct reals for which we have equality in (5.1). The complexity of the proof of this result is of a different order than that of the Erdős-Szekeres theorem, although there are important qualitative insights that Chung's proof shares with the dynamic programming of the Erdős-Szekeres theorem that was given in Section two. The proof provided by Chung calls instead on four functions parallel to the two components φ of Section two, although instead of considering only one simple injective image contained in $[1, n] \times [1, n]$, Chung must consider several such images that are contained in more complex domains.

With the deterministic problem resolved, Chung (1980) posed the natural problem of determining the asymptotic behavior of $U_n(X_1, X_2, \ldots, X_n)$ when the X_i are independent and random variables with a common continuous distribution. This turned out to be much easier to resolve than the problem solved by Chung, and in Steele (1981) it was proved that we have that

$$U_n/\sqrt{n} \to 2\sqrt{2}$$

with probability one. The corresponding constant when one permits k changes in the sense of the monotonicity was also found to be $2\sqrt{k}$.

6. Concentration inequalities. The length I_n of the longest increasing subsequence of n independent uniformly distributed random variables has the property of being rather tightly concentrated about its mean. One way to see this in terms of variance is to call on the Efron-Stein inequality which in the form given by Steele (1986) tells us that if $F(y_1, y_2, \ldots, y_{n-1})$

is any function of $n-1$ variables then by introducing n random variables by applying F to the variables $X_1, X_2, \ldots, X_n$ with the i'th variable withheld $F_i = F(X_1, X_2 \ldots, X_{i-1}, X_{i+1}, \ldots, X_n)$ and setting $\overline{F} = \frac{1}{n}(F_1 + F_2 + \ldots + F_n)$, we have

$$\operatorname{Var}\left(F(X_1, X_2, \ldots, X_{n-1})\right) \leq E \sum_{i=1}^{n} (F_i - \overline{F})^2.$$

When we focus on I_{n-1} we first note that

$$I(X_1, X_2, \ldots, X_n) \leq 1 + I(X_1, X_2, \ldots, X_{i-1}, X_{i+1}, \ldots X_n)$$

and

$$I(X_1, X_2, \ldots X_{i-1}, X_{i+1}, \ldots X_n) \leq I(X_1, X_2, \ldots, X_n).$$

If we let A_n denote the number of sample points X_i that are in *all* of the increasing subsequence having maximum length L_n, we then find

$$\sum_{i=1}^{n} \left(I(X_1, X_2, \ldots, X_n) - I(X_1, X_2, \ldots, X_{i-1}, X_{i+1}, \ldots, X_n)\right)^2 \leq A_n$$

Since $A_n \leq I_n$ and since the quadratic sum is decreased by replacing $I(X_1, X_2, \ldots, X_n)$ by $n^{-1}(I(X_2, X_3, \ldots, X_n) + I(X_1, X_2, \ldots, X_n) + \ldots + I(X_1, X_2, \ldots, X_{n-1}) = \overline{I}$ we have

$$\sum_{i=1}^{n} (I(X_1, X_2, \ldots, X_{i-1}, X_{i+2}, \ldots X_n) - \overline{I})^2 \leq I_n.$$

Taking expectations and applying the Efron-Stein inequality gives

$$\operatorname{Var} I_{n-1} \leq E I_n \leq C \sqrt{n}$$

where for any $n \geq n(\varepsilon)$ we can take $C \leq 2 + \varepsilon$.

This bound on $\operatorname{Var} I_{n-1}$ was easily won, but despite its simplicity is enables us to circumvent some rather heavy analysis. In particular, the variance bound, monotonicity of I_n, and the asymptotics of the mean $E I_n \sim 2\sqrt{n}$ give us an easy proof that $I_n / \sqrt{n} \to 2$ with probability one just by following the usual Chebyshev and subsequence arguments. This development may seem surprising since the strong law for I_n has served on several occasions as a key example of the effectiveness of subadditive ergodic theory (Durrett (1991) and Kingman (1973)). Here we should also note that Aldous (1993) gives a scandalously simple proof of the Efron-Stein inequality, making it easy enough to cover in even a first course in probability theory.

Lower bounds on Var I_n have been obtained recently by B. Bollobás and R. Pemantle who independently established that there is a $c > 0$ for which

$$\text{Var } I_n \geq cn^{1/8}$$

for all $n \geq 3$. In view of the bounds just reviewed one suspects that

$$\lim_{n \to \infty} \log(\text{Var } I_n)/\log n = \alpha$$

for some $1/8 \leq \alpha \leq 1/2$. The existence of the limit may not be too difficult to establish, but substantial new insight will be required to determine the value of α.

The tail probabilities of I_n were first studied in Frieze (1991) using the bounded difference method. The bound obtained by Frieze was subsequently improved by Bollobás and Brightwell (1992) who established that for all $\epsilon > 0$ there is a $\beta = \beta(\epsilon) > 0$ such that for $n \geq n(\varepsilon)$, we have

$$P\left(|I_n - EI_n| \geq n^{1/4+\varepsilon}\right) \leq \exp(-n^\beta).$$

The work of Bollobás and Brightwell (1992) also considered the d-dimensional increasing subsequences $\Lambda_d^+(X_1, X_2, \ldots, X_n) = \Lambda_{d,n}^+$ for X_i independent and uniformly distributed on the unit d-cube.

THEOREM 6.1. *For every $d \geq 2$, there is a constant A_d such that for all $n \geq n(d)$ one has*

$$P\left(|\Lambda_{d,n}^+ - E\Lambda_{d,n}^+| \geq \lambda A_d n^{1/2d} \log n/\log\log n\right) \leq 80\lambda^2 \exp(-\lambda^2)$$

for all λ with $2 < \lambda < n^{1/2d}/\log\log n$.

Large deviation results like the last one have many consequences but a particularly valuable consequence in the present case is that one can extract a rate of convergence result for certain means. We let N have the Poisson distribution with mean n and define constants $c_{n,d}$ by

$$n^{-1/d}E\Lambda_d^+(X_1, X_2, \ldots, X_N) = c_{n,d}.$$

The main result is that

$$c_d - c_{n,d} = 0\left(n^{-1/2d}\log^{3/2} n/\log\log n\right)$$

where $c_2 = 2$ and the other constants c_d are those of the Bollobás-Winkler theorem for $d > 2$ as viewed in Section 3.

The most precise result on the deviations of the length longest increasing subsequence I_n is due to Talagrand (1993) and emerges from the theory of abstract isoperimetric theorems for product measures developed in Talagrand (1988, 1991, 1993).

THEOREM 6.2. *If M_n denotes the median of I_n, then for all $u > 0$ we have*

$$P(I_n \geq M_n + u) \leq 2\exp(-u^2/(4(M_n - u)))$$

and

$$P(I_n \leq M_n - u) \leq 2\exp(-u^2/(4M_n)).$$

7. Pseudo-random sequences. If $0 < \alpha < 1$ is an irrational number and $\{X\}$ denotes the fractional part of X, then the sequence of values determined by the fractional parts of the multiples of α given by $\{\alpha\}, \{2\alpha\}, \{3\alpha\}, \ldots, \{n_\alpha\}$ are uniformly distributed in $[0, 1]$ in the sense that if $[a, b] \subset [0, 1]$ then the number of the integers k satisfying $\{k\alpha\} \in [a, b]$ for $1 \leq k \leq n$ is asymptotic to $(b - a)n$ as $n \to \infty$. Bohl (1909) was evidently the first to show sequence of points $\{n\alpha\}$ share this property with the independent uniformly distributed random variables X_n, $1 \leq n < \infty$, though the subsequent importance of this observation was certainly driven home by H. Weyl and G. Hardy. The project of exploring which properties of the X_n are shared by the $\{n\alpha\}$ is a natural one, of which the results of Kesten (1960) and Beck (1991) are telling examples. The survey of Niederreiter (1978) provides an extensive review of the ways that pseudo-random numbers can parallel those that are honestly random and also articulates ways where pseudo-random sequences can be even more useful.

There are some very simple properties that make $\{X_n\}$ and $\{n\alpha\}$ seem quite different, so there is particular charm to the fact that they can behave quite similarly in the context of such a non-standard issue as the length of the longest increasing subsequence. If we let $\ell_n^+(\alpha)$ and $\ell_n^-(\alpha)$ denote respectively the longest increasing and longest decreasing subsequences of $\{\alpha\}, \{2\alpha\}, \ldots, \{n\alpha\}$, then ℓ_n^+ and ℓ_n^- turn out to have behavior that echos closely the behavior of their stochastic cousins. The first investigation of this issue is given in Del Junco and Steele (1978) where it is proved using discrepancy estimates that

$$\frac{\log \ell_n^+(\alpha)}{\log n} \to \frac{1}{2} \text{ and } \frac{\log \ell_n^-(\alpha)}{\log n} \to \frac{1}{2}$$

for all algebraic irrationals α and for a set of irrationals of $[0, 1]$ measure one.

In Boyd and Steele (1978) a more precise understanding of $\ell_n^+(\alpha)$ was obtained by using the continued fraction expansion of α. It turns out that $n^{1/2}$ is the correct order of ℓ_n^+ and ℓ_n^- if and only if α has bounded partial quotients. Moreover, when the partial quotient sequence of α is known one can determine the precise range of $\ell_n^+/\sqrt{n}$ and $\ell_n^-/\sqrt{n}$. For example, in the eternally favorite special case of the golden ratio $\alpha_0 = (1 + \sqrt{5})/2 = [1; 1, 1, 1, \ldots]$, we have

$$\liminf_{n \to \infty} \ell_n^+(\alpha_0)/\sqrt{n} = 2/5^{1/4} = 1.337481\ldots$$

and

$$\limsup_{n \to \infty} \ell_n^-(\alpha_0)\sqrt{n} = 5^{1/4} = 1.495349\ldots .$$

There is a non-zero gap $\Delta(\alpha_0)$ between these two limits, and in fact there is no α for which one has a limit theorem precisely like that for independent random variables, but nevertheless we see there is a close connection between $\{n\alpha\}$ and the genuinely random case.

In Boyd and Steele (1978) it is further proved that the gap $\Delta(\alpha)$ is minimized precisely for $\alpha_0 = (1 + \sqrt{5})/2$. It is also proved there that for all irrational α we have

$$\limsup_{n \to \infty} \ \ell_n^+(\alpha)\ell_n^-(\alpha)/n = 2,$$

and for α with unbounded partial quotients

$$\liminf_{n \to \infty} \ \ell_n^+(\alpha)\ell_n^-(\alpha)/n = 1.$$

These results contrast with the analog for a sequence of independent uniformly distributed random variables for which we have

$$\lim_{n \to \infty} \ I_n^+ I_n^- = 4 \quad \text{a.s.}$$

The behavior of the pseudo-random Weyl sequence $\{\alpha\}$, $\{2\alpha\}$, $\{3\alpha\},\ldots$, $\{n\alpha\}$ is not unique in the world of pseudo-random sequences. There is a remarkable sequence due to van der Corput (cf. Hammersley and Handscomb (1964)) that was invented in order to provide a sequence $\{x_i \in [0,1] : 1 \le i < \infty\}$ that has an especially small discrepancy

$$D_n = \sup_{0 \le x \le 1} |\sum_{i=1}^{n} 1_{[0,x]}(x_i) - x|.$$

To define this sequence we first note that for any integer $n \ge 0$ there is a unique representation $n = \sum_{i=0}^{\infty} a_i 2^i$ where $a_i \in \{0,1\}$. The n-th element $\varphi_2(n)$ of the van der Corput sequence of base 2 is given by "reflecting the expansion of n in the decimal point," that is we have

$$\varphi_2(n) = \sum_{i=0}^{\infty} a_i 2^{-i-1}.$$

The sequence is thus given by $\{1/2, 1/4, 3/4, 1/8, 5/8, 3/8, 7/8, \ldots\}$, and one can see that the sequence does indeed disperse itself in a charmingly uniform fashion. Béjian and Faure (1977) have shown that there are sequences that are still more uniform but even there new sequences are conceptually quite close to the original idea.

Asymptotic equidistribution is perhaps the most basic feature of a sequence of independent uniformly distributed sequence, but one can check that the van der Corput sequence is far more uniform than a random sequence. This divergence from the behavior of independent uniform random variables is one of the facts that adds zest to the study of the behavior of the longest increasing subsequence of $\{\varphi_2(k) : 1 \leq k \leq n\}$.

The basic results are the following:

$$\limsup_{n \to \infty} \ell^+(\varphi_2(1), \varphi_2(2), \dots, \varphi_2(n))/\sqrt{n} = 3/2$$

and

$$\liminf_{n \to \infty} \ell^+(\varphi_2(1), \varphi_2(2), \dots, \varphi_2(n))/\sqrt{n} = \sqrt{2}.$$

For any integer p, one can define a *base* p van der Corput sequence by expanding n in base p and letting $\varphi_p(n)$ be the analogous "reflection in the decimal point" for $p > 2$ we have

$$\limsup_{n \to \infty} \ell^+(\varphi_p(1), \varphi_p(2), \dots, \varphi_p(n))/\sqrt{n} = p^{1/2}$$

and

$$\liminf_{n \to \infty} \ell^+(\varphi_p(1), \varphi_p(2), \dots, \varphi_p(n))/\sqrt{n} = 2(1 - p^{-1})^{1/2}.$$

Because of the interest that has been devoted to the hard won constant $c_2 = 2$ for the limit of $\ell_n^+/\sqrt{n}$ for the case of independent uniform variables, we should record that the last limit tells us that

$$\lim_{p \to \infty} \liminf_{n \to \infty} \ell^+(\varphi_p(1), \varphi_p(2), \dots, \varphi_p(n))/\sqrt{n} = 2.$$

8. Theory of subsequences of sections. Many combinatorial problems exhibit new behaviors when they are imbedded in an infinite sequence of nested problems. The most classical instance of this phenomenon is offered by the theory of irregularity of distribution (cf. Beck and Chen (1987)).

To see how this phenomenon is manifested in the context of the Erdős-Szekeres theorem we suppose we are given an infinite sequence of distinct reals $S = \{x_1, x_2, \dots, x_n, \dots\}$, we focus on $M(x_1, x_2, \dots, x_n) = M_n(S)$ which denote the cardinality of the largest monotone subsequence of $\{x_1, x_2, \dots, x_n\}$. The Erdős-Szekeres theorem tells us that $M_n \geq \sqrt{n}$ for all $n \geq 1$, and, this is the best one can say for any fixed n. Still, for any infinite sequence S of distinct reals we can do better than $\sqrt{n}$ on infinitely many blocks of length n.

THEOREM 8.1. *There is a constant $\gamma > 1$ such that*

$$(8.1) \qquad \limsup_{i, n \to \infty} M(x_{i+1}, x_{i+2}, \dots, x_{i+n})/\sqrt{n} \geq \gamma.$$

Before embarking on the proof of this result we provide an example that shows that a doubly indexed result is what one required if the intention is to beat the Erdős-Szekeres theorem by a factor exceeding 1. We can construct an infinite sequence of integers $\{x_1, x_2, \ldots\}$ such that for all of the finite sections $\{x_1, x_2, \ldots, x_n\}$ we have $M(x_1, x_2, \ldots, x_n) = \lceil \sqrt{n} \rceil$. This sequence can be constructed as a concatenation of blocks

$$B_k = \{(-1)^{k+j} 3^k + (-1)^{k+j}(2k - j) : j = 0, 1, 2, \ldots, 2k\}$$

for $k = 0, 1, 2, \ldots$. We note that $|B_k| = 2k + 1$, so $|B_0| + |B_1| + \ldots + |B_k| = (k + 1)^2$. Also, we note that $B_0 = \{0\}$, $B_1 = \{-5, 4, -3\}$, $B_2 = \{13, -12, 11, -10, 9\}$ and $B_3 = \{-33, 32, -31, 30, -29, 28, -27\}$.

The proof of Theorem 8.1 depends on results that tell us that if $M(x_1, x_2, \ldots, x_n)$ and $M(x_1, x_2, \ldots, x_{2n})$ are both small, then $M(x_{n+1}, x_{n+2}, \ldots, x_{2n})$ must be exceptionally large. We will provide one such result as a consequence of the next lemma, but we first require some notation. Given any fixed sequence S, we let

$$a(k) = \max\{t : \exists x_{i_1} > x_{i_2} > \ldots > x_{i_t} \text{ with } 1 \le i_1 < i_2 < \ldots < i_t = k\}$$
$$b(k) = \max\{t : \exists x_{i_1} < x_{i_2} < \ldots < x_{i_t} \text{ with } 1 \le i_1 < i_2 < \ldots < i_t = k\}$$
$$c_n(k) = \max\{t : \exists x_{i_1} > x_{i_2} > \ldots > x_{i_t} \text{ with } k = i_1 < i_2 < \ldots < i_t \le n\}$$
$$d_n(k) = \max\{t : \exists x_{i_1} < x_{i_2} < \ldots < x_{i_t} \text{ with } k = i_1 < i_2 < \ldots < i_{i_t} \le n\}.$$

To illuminate these definitions, we note $a(k)$ is the length of the longest decreasing sequence terminating with x_k, and $c_n(k)$ is the length of the longest decreasing subsequence beginning with x_k and ending before the nth element of the sequence. Thus, the sum $a(k) + c_n(k)$ is the length of the longest decreasing subsequence of $\{x_1, x_2, \ldots, x_n\}$ that contains x_k.

LEMMA 8.1. *For any $0 < \varepsilon < 1/2$ and $n \ge n_0(\varepsilon)$, if we have the bound $M(x_1, x_2, \ldots x_n) \le (1 + \varepsilon)\sqrt{n}$, then there exists $1 \le k \le n$ such that*

$$a(k) \ge (1 - \delta)\sqrt{n}$$

and

$$b(k) \ge (1 - \delta)\sqrt{n}$$

provided that $\delta^2 > 2\varepsilon$.

Proof. The mapping $k \to (a(k), b(k))$ is injective since if $k < k'$ then we have $a(k) < a(k')$ or $b(k) < b(k')$, or both, because $x_{k'}$ will extend at least one of the monotone sequences ending at x_k. The set of integers (s, t) with $1 \le s, t \le (1 + \varepsilon)\sqrt{n}$ and $\min(s, t) < (1 - \delta)\sqrt{n}$ has cardinality at most $(1 - \delta)^2 n + 2(\varepsilon + \delta)(1 - \delta)n < (1 - \delta^2 + 2\varepsilon)n$ and the n distinct points $(a(k), b(k))$ are among the lattice points $[1, M_n]^2 \subset [1, (1 + \varepsilon)\sqrt{n}]^2$, so for $\delta^2 > 2\varepsilon$ the pigeonhole principle yields the lemma. $\square$

PROPOSITION 8.1. *For $0 < \varepsilon < 1/2$ and all $n \geq n_0(\varepsilon)$, the inequalities*

$$M(x_1, x_2, \ldots, x_n) \leq (1+\varepsilon)\sqrt{n}$$
$$M(x_1, x_2, \ldots, x_{2n}) \leq (1+\varepsilon)\sqrt{2n}$$

imply

$$M(x_{n+1}, x_{n+2}, \ldots, x_{2n}) \geq (1 - 25\delta)\sqrt{2n}$$

provided $\delta^2 > 2\varepsilon$.

Proof. Even before beginning, we note that the factor of 25 given above can be improved, but it suffices for our main point and allows for simple computations. By the preceding lemma we have $1 \leq k \leq n$ such that

$$a(k) \geq (1 - \delta)\sqrt{n} \text{ and } b(k) \geq (1 - \delta)\sqrt{n}.$$

By considering subsequences that go through x_k and continue from x_j with $n \leq j \leq 2n$ we see that

$$M(x_1, x_2, \ldots, x_{2n}) \geq (1 - \delta)\sqrt{n} + \min\{c_{2n}(j), d_{2n}(j)\},$$

so, by the bound on $M(x_1, x_2, \ldots, x_{2n})$, we find

(8.2) $$\min\{c_{2n}(j), d_{2n}(j)\} < (1+\varepsilon)\sqrt{2n} - (1-\delta)\sqrt{n}$$

for all $n < j \leq 2n$. Now, unless the conclusion of the proposition holds we also have

(8.3) $$\max\{c_{2n}(j), d_{2n}(j)\} \leq (1 - 25\delta)\sqrt{2n},$$

so the issue is to estimate the number of $n < j \leq 2n$ that can satisfy (5.2) and (5.3). Since the mapping $g \mapsto (c_{2n}(j), d_{2n}(j))$ is injective, the proposition follows if there are fewer than n solutions of (5.2) and (5.3).

We need to count the number of positive lattice points (i, j) with

$$\min\{i, j\} < \{(1 + \varepsilon)\sqrt{2} - 1 + \delta\}\sqrt{n}$$

and

$$\max\{i, j\} < \sqrt{2}(1 - 25\delta)\sqrt{n}.$$

A calculation shows that the area of the L-shaped region defined by these inequalities is bounded by $n\{1 - \delta\} < n$. $\square$

To complete the proof of the main theorem of this section, we first note that setting $\delta^2 = 2\varepsilon$ and solving $1 + \varepsilon = (1 - 25\delta0\sqrt{2}$ we are lead to a value of $\delta = 0.117118$. This tells us that in the theorem we can take any $j \leq 1 + \delta^2/2 \leq 1.00014$.

The key problem that remains open at this point is the determination of the best possible value of j.

9. Subsequences along cycles. We say that a sequence of integers $(i_1, i_2, \ldots, i_k)$ has d *descents* if it can be written as the concatination of d monotonic blocks but cannot be written as a concatination of fewer than d monotonic blocks.

For example, the sequence $(5, 9, 11, 2, 3, 8, 4, 1)$ can be written as the concatination of four monotonic increasing blocks $(5, 9, 11)$, $(2, 3, 8)$, (4), (1) and thus the sequence has $d = 4$ descents. We let

$$\ell^+(d; x_1, x_2, \ldots, x_n) =$$

$$\max\{k : x_{i_1} < x_{i_2} < \ldots < x_{i_k} \text{ where } (i_1, i_2, \ldots, i_k) \text{ has } d \text{ descents}\}.$$

We define $\ell^-(d; x_1, x_2, \ldots, x_n)$ as the corresponding maximum length d-descent monotone sequence of x_i's. The purpose of introducing these quantities is that they lead to a quite natural analog of the Erdős-Szekeres theorem.

THEOREM 9.1. *For any n distinct real numbers, we have*

$$\ell^+(d; x_1, x_2, \ldots, x_n)\ell^-(d; x_1, x_2, \ldots, x_n) \geq dn.$$

Proof. We first remark that the case $d = 1$ is the usual Erdős-Szekeres theorem. Further in the trivial case $d = n$, we have equality since $\ell^+ = \ell^- = n$. □

10. Common ascending subsequences. If π and σ are two permutations of $\{1, 2, \ldots, n\}$, we say they have a *common ascending subsequence of length r* if there are indices $1 \leq i_1 < i_2 < \ldots < i_r \leq n$ and $1 \leq j_1 < j_2 < \ldots < j_r \leq n$ such that $\pi(i_s) = \sigma(j_s)$ for all $1 \leq s \leq r$. The notion of the longest common ascending subsequence $\lambda(\pi, \sigma)$ was introduced in Alon (1990) for the purpose of derandomizing the randomized maximum flow algorithm of Cheriyan and Hagerup (1989). The connection between $\lambda(\pi, \sigma)$ and the central theme of this review is made most explicit by noting that $\lambda(\pi, \sigma)$ is equal to the length of the longest increasing subsequence of $\sigma^{-1}\pi(1), \sigma^{-1}\pi(2), \ldots, \sigma^{-1}\pi(n)$. The main result of Alon (1990) is the following theorem.

THEOREM 10.1. *For every two integers k and n with $k \geq n$, one can construct in time $0(kn)$, a sequence of permutations $\pi_1, \pi_2, \ldots, \pi_n$ of $\{1, 2, \ldots, n\}$ such that any permutation σ satisfies*

$$\frac{1}{k} \sum_{i=1}^{k} \lambda(\sigma, \pi_i) = 0(n^{2/3}).$$

The fact that this result is reasonably sharp can be deduced from the probabilistic results of the first section. For any fixed $\pi_1, \pi_2, \ldots, \pi_n$, we have for a random σ that

$$E\frac{1}{k} \sum_{i=1}^{k} \pi(\sigma, \pi_i) = EI_n \geq (2 - \varepsilon)\sqrt{n}$$

for any $\varepsilon > 0$ and $n \geq n(\varepsilon)$. Combining this observation with Alon's theorem with $k = n$ we find that there is a constant $c > 0$ such that the functional

$$A_n = \max_\sigma \min_{\{\pi_i\}} \frac{1}{n} \sum_{i=1}^{n} \lambda(\sigma, \pi_i)$$

satisfies

$$(2 - \varepsilon)\sqrt{n} \leq A_n \leq cn^{2/3}.$$

The main point of reviewing this information is to point out the problem of determining the true order of A_n.

11. Optimal sequential selection. One of the intriguing themes of sequential selection is that one sometimes does surprisingly well in making selections even without knowledge of the future or recourse to change past choices. One illustration of the theme is the "secretary problem" of Gilbert and Mosteller (1966) that tells us that given $X_1, X_2, \ldots, X_n$ independent and identically distributed random variables, there is a stopping time $\tau = \tau_n$ such that

$$P(X_\tau = \max_{1 \leq i \leq n} X_i) > e^{-1}.$$

There turns out to be a parallel phenomenon that takes place in the theory of monotone subsequences.

The problem is to determine how well one can make a sequence of choices from a sequentially reveal set of independent random variables with a known common continuous distribution. To put this problem rigorously, we call a sequence of stopping times $1 \leq \tau_1 < \tau_2 < \ldots$ a *policy* if they are adapted to $X_1, X_2, \ldots$ and if we have $X_{\tau_1} < X_{\tau_2} < \ldots < X_{\tau_k} < \ldots$. We let $\mathcal{S}$ denote the set of all policies. The main result about such policies is given in Samuels and Steele (1981).

THEOREM 11.1. *For any sequence of independent random variables with continuous distribution F and associated set of policies $\mathcal{S}$, we have*

$$u_n = \sup_\mathcal{S} E(\max\{k : \tau_k \leq n\}) \sim \sqrt{2n}$$

as $n \to \infty$.

This theorem thus tells us that there is a policy by which we can make sequential selections from $X_1, X_2, \ldots, X_n$ and obtain a monotone increasing subsequence of length that is asymptotic in expectation to $\sqrt{2n}$. Since the best one could do with full knowledge of $\{X_1, X_2, \ldots, X_n\}$ is to obtain a subsequence with length that is asymptotic in expectation to $2\sqrt{n}$, we see that a "mortal" does worse than a "prophet" by only a factor of $\sqrt{2}$.

It was observed by Burgess Davis (cf. Samuels and Steele (1981), Section 7), that if one lets ℓ_n denote the expected value of length of the longest increasing subsequence that can be made sequentially from a random permutation of $\{1, 2, \ldots, \}$, then one has $\ell_n \sim u_n$ and consequently $\ell_n \sim \sqrt{2n}$. The importance of this remark comes from the fact that the "natural" selection studied in Baer and Brock (1968) is just "sequential" selection as studied here. Thus, the results of Samuels and Steele (1981) coupled with the key observation of Burgess Davis resolve the main problem posed in Baer and Brock (1968).

12. Open problems and concluding remarks. Erdős (see Chung (1980), p. 278) has raised the question of determining the optimum values associated with various *weighted* versions of the monotone and k-modal subsequence problems. Specifically, let $\mathcal{W} = \{(w_1, w_2, \ldots w_n) \; : \; w_i \geq 0 \text{ and } w_1 + w_2 + \ldots + w_n = 1\}$ and for $w \in \mathcal{W}$ and for $0 \leq k$ let $\mathcal{U}(w)$ denote the set of k-unimodal subsequences of w. The problem is to determine the values

$$\tau(n, k) = \min_{w \in \mathcal{W}} \max_{u \in \mathcal{U}(w)} \sum_{w_i \in u} w_i.$$

To illustrate, we note that it is easy to show that $\tau(n, 0) \leq n^{1/2}$, since by considering a perturbation of uniform weights that have the same ordering which yields a longest monotone subsequence of length $\lceil u^{1/2} \rceil$, we see $\tau(n, 0) \leq n^{-1} \lceil n^{1/2} \rceil$.

One surely suspects that $\tau(n, 0)\sqrt{n} \to 1$ as $n \to \infty$, but this has not yet been established, though it might be easy. By similar considerations using Chung's theorem, one sees that $\limsup \tau(n, 1)\sqrt{n} \leq \sqrt{3}$ while we expect that actually $\tau(n, 1) \to \sqrt{3}$.

Erdős posed the problem of determining the largest integer $f(n)$ such that any sequence of $m = f(n)$ distinct real numbers $x_1, x_2, \ldots, x_m$ can be decomposed into n monotonic sequences. Hanani (1957) proved that

$$f(n) = n(n + 3)/2.$$

A question posed by Erdős (1973) for which there seems to have been no progress is the following:

Given $x_1, x_2, \ldots, x_n$ distinct real numbers determine

$$\max_M \sum_{i \in M} x_i$$

where the maximum is over all subsets of indeces $1 \leq i_1 < i_2 < \ldots < i_k \leq n$ such that $x_{i_1}, x_{i_2}, \ldots, x_{i_k}$ is monotone.

Acknowledgement. I am pleased to thank D. Aldous, P. Diaconis and M. Talagrand for making available prepublication copies of their work. I also thank L. Lovász for comments that contributed to the discussion in Section nine, S. Janson for comments that contributed to Section six, and P. Winkler for comments that contributed to Section three. P. Erdős provided the reference to the work of Hanani, as well as the inspiration to pursue this survey.

REFERENCES

[1] D. ALDOUS, *Reversible Markov chains and random walks on graphs*, Monograph Draft, version of January 25, 1993, Department of Statistics, University of California, Berkeley, CA (1993).

[2] D. ALDOUS AND P. DIACONIS, *Hammersley's interacting particle process and longest increasing subsequences*, Technical Report, Department of Statistics, U.C. Berkeley, Berkeley, CA (1993).

[3] N. ALON, *Generating pseudo-random permutations and maximum flow algorithms*, Information Processing Letters 35 (1990), pp. 201–204.

[4] N. ALON, B. BOLLOBÁS, G. BRIGHTWELL AND S. JANSON, *Linear extensions of a random partial order*, to appear in Ann. Appl. Prob. 3 (1993).

[5] R.M. BAER AND P. BROCK, *Natural sorting over permutation spaces*, Math. Comp. 22 (1968), pp. 385–510.

[6] J. BECK, *Randomness of $n\sqrt{2} \bmod 1$ and a Ramsey property of the hyperbola*, Colloquia Mathematica Societatis János Bolyai 60 Sets, Graphs, and Numbers, Budapest, Hungary (1991).

[7] J. BECK AND W.W.L. CHEN, *Irregularities of Distribution*, Cambridge University Press, 1987.

[8] R. BÉJIAN AND H. FAURE, *Discrépance de la suite de van der Corput*, C.R. Acad. Sci. Paris Sér. A 285 (1977), pp. 313–316.

[9] P. BLACKWELL, *An alternative proof of a theorem of Erdös and Szekeres*, Amer. Mathematical Monthly 78 (1971), p. 273.

[10] K.P. BOGART, C. GREENE, AND J.P. KUNG, *The impact of the chain decomposition theorem on classical combinatorics*, in The Dilworth Theorems: Selected Papers of Robert P. Dilworth (K.P. Bogart, R. Freese, and J.P. Kung, eds.), Birkhäuser Publishers, Boston (1990), pp. 19–29.

[11] P. BOHL, *Über ein in der Theorie der säkularen Störunger vorkommendes Problem*, J. Reine U. Angew. Math. 135 (1909), pp. 189–283.

[12] B. BOLLOBÁS AND G. BRIGHTWELL, *The height of a random partial order: Concentration of measure*, Ann. Appl. Prob. 2 (1992), pp. 1009–1018.

[13] B. BOLLOBÁS AND P.M. WINKLER, *The longest chain among random points in Euclidean space*, Proc. Amer. Math. Soc. 103 (1988), pp. 347–353.

[14] G. BRIGHTWELL, *Models of Random Partial Orders in Surveys in Combinatorics*, 1993, London Mathematical Society Lecture Notes 187 (K. Walker, ed.), (1993).

[15] J. CHERIYAN AND T. HAGERUP, *A randomized maximum-flow algorithm*, Proceedings of the IEEE-FOCS 1989 (1989), pp. 118–123.

[16] F.P.K. CHUNG, *On Unimodal Subsequences*, J. Combinatorial Theory (A) 29 (1980), pp. 267–279.

[17] J.G. VAN DER CORPUT, *Verteilungs funktionen, I,II*, Nederl. Akad. Wetensch. Proc. 38 (1935), pp. 813–821, 1058–1066.

[18] A. DEL JUNCO AND J.M. STEELE, *Hammersley's Law for the van der Corput Sequence: An Instance of Probability Theory for Pseudorandom Numbers*, Annals of Probability 7 (2) (1979), pp. 267–275.

[19] R. DURRETT, *Probability: Theory and examples*, Wadsworth, Brooks/Cole, Pacific Grove, CA (1991).

[20] R.P. DILWORTH, *A Decomposition Theorem for Partially Ordered Sets*, Annals of Mathematics 51 (1950), pp. 161–165.

[21] P. ERDÖS, KO CHAO, AND R. RADO, *Intersection theorems for systems of finite sets*, Quarterly J. Mathematics Oxford Series (2) 12 (1961), pp. 313–320.

[22] P. ERDÖS, *Unsolved problems in Graph Theory and Combinatorial Analysis*, Combinatorial Mathematics and It's Applications (Proc. Conf. Oxford 1969), Academic Press, London (1971), pp. 97–109. Also in *Paul Erdös: The Art of Counting*, (J. Spencer, ed.), MIT Press, Cambridge, MA (1973).

[23] P. ERDÖS AND R. RADO, *Combinatorial theorems on classifications of subsets of a given set*, Proc. London Math Soc. 3 (2) (1952), pp. 417–439.

[24] P. ERDÖS AND GY. SZEKERES, *A combinatorial problem in geometry*, Compositio Math. 2 (1935), pp. 463–470.

[25] A.M. FRIEZE, *On the length of the longest monotone subsequence in a random permutation*, Ann. Appl. Prob. 1 (1991), pp. 301–305.

[26] T. GALLAI AND N. MILGRAM, *Verallgemeinerung eines graphentheoretischen Satzes von Reide*, Acta Sci. Math. 21 (1960), pp. 181–186.

[27] J.P. GILBERT, AND F. MOSTELLER, *Recognizing the maximum of a sequence*, J. Amer. Statist. Soc. 61 (1966), pp. 228–249.

[28] J.M. HAMMERSLEY, *A few seedlings of research*, Proc. 6th Berkeley Symp. Math. Stat. Prob., U. of California Press (1972), pp. 345–394.

[29] J.M. HAMMERSLEY AND D.C. HANDSCOMB, *Monte Carlo Methods*, Methuen Publishers, London (1964).

[30] H. HANANI, *On the number of monotonic subsequence*, Bull. Res. Council Israel, Sec. F (1957), pp. 11–13.

[31] H. KESTEN, *Uniform distribution* mod 1, Annals of Math. 71 (1960), pp. 445–471.

[32] J.F.C. KINGMAN, *Subadditive ergodic theory*, Ann. Prob. 1 (1973), pp. 883–909.

[33] D. KNUTH, *Permutations, matrices, and generalized Young tableaux*, Pacific J. Mathematics 34 (1970), pp. 709–727.

[34] W. KRASKIEWICZ, *Reduced decompositions in hyperoctahedral groups*, C.R. Acad. Sci. Paris 309 (1989), pp. 903–907.

[35] V. LIFSCHITZ AND B. PITTEL, *The number of increasing subsequences of the random permutation*, J. Comb. Theory Ser. A 31 (1981), pp. 1–20.

[36] B.F. LOGAN AND L.A. SHEPP, *A variational problem for Young tableaux*, Advances in Mathematics 26 (1977), pp. 206–222.

[37] L. LOVÁSZ, *Combinatorial Problems and Exercises*, North-Holland Publishing Company, Amsterdam (1979).

[38] H. NIEDERREITER, *Quasi-Monte Carlo methods and pseudo-random numbers*, Bull. Amer. Math. Soc. 84 (1978), pp. 957–1041

[39] S. PILPEL, *Descending subsequences of random permutations*, J. Comb. Theory, Ser. A 53 (1) (1990), pp. 96–116.

[40] S. SAMUELS AND J.M. STEELE, *Optimal Sequential Selection of a Monotone Subsequence from a random sample*, Ann. Prob. 9 (1981), 937–947.

[41] C. SCHENSTED, *Longest increasing and decreasing subsequences*, Canadian J. Math. 13 (1961), pp. 179–191.

[42] A. SEIDENBERG, *A simple proof of a theorem of Erdös and Szekeres*, J. London Math. Soc. 34 (1959), p. 352.

[43] R.S. SILVERSTEIN, *Estimation of a constant in the theory of three dimensional random orders*, BS/MS Thesis in Mathematics (1988), Emory University.

[44] D. STANTON, AND D. WHITE, *Constructive Combinatorics*, Springer-Verlag, New York (1986).

[45] J.M. STEELE, *Limit properties of random variables associated with a partial ordering of R^d*, Ann. Prob. 5 (1977), pp. 395–403.

[46] J.M. STEELE, *Long unimodal subsequences: A problem of F.R.K. Chung*, Discrete Mathematics 33 (1981), pp. 223–225.

[47] J.M. STEELE, *An Efron-Stein inequality for non-symmetric statistics*, Annals of Statistics 14 (1986), pp. 753–758.

[48] M. TALAGRAND, *An isoperimetric theorem on the cube and the Khintchin-Kahane inequalities*, Proc. Amer. Math. Soc. 104 (1988), pp. 905–909.

[49] M. TALAGRAND, *A new isoperimetric inequality for product measure and the tails of sums of independent random variables*, Geometric and Functional Analysis, 1 (2) (1991), pp. 211–223.

[50] M. TALAGRAND, *Concentration of measure and isoperimetric inequalities in a product space*, Technical Report, Department of Mathematics, Ohio State University, Columbus, OH (1993).

[51] W.T. TROTTER, *Combinatorics and Partially Ordered Sets: Dimension Theory*, The Johns Hopkins University Press, Baltimore (1992), pp. xiv + 307.

[52] S.M. ULAM, *Monte Carlo calculations in problems of mathematical physics*, in Modern Mathematics for the Engineer (E.F. Beckenbach, ed.) McGraw-Hill, New York (1961).

[53] A.M. VERŠIK AND S.V. KEROV, *Asymptotics of the Plancherel measure of the symmetric group and the limiting form of Young tableaux*, Dokl. Akad. Nauk. SSSR 233 (1977), pp. 1024–1028.

[54] A.M. VERŠIK AND S.V. KEROV, *Asymptotics of the largest and the typical dimensions of irreducible representations of the symmetric group*, Functional Analysis and Its Applications 19 (1985), pp. 21–31.

[55] D.B. WEST, *Extremal Problems in partially ordered sets*, in Ordered Sets: Proceedings of the Banff Conference (I. Rival, eds.) Reidel, Dordrecht and Boston (1982), pp. 473–521.

RANDOMISED APPROXIMATION SCHEMES FOR TUTTE-GRÖTHENDIECK INVARIANTS

DOMINIC WELSH*

1. Introduction. Consider the following very simple counting problems associated with a graph G.

(i) What is the number of connected subgraphs of G?

(ii) How many subgraphs of G are forests?

(iii) How many acyclic orientations has G?

Each of these is a special case of the general problem of evaluating the Tutte polynomial of a graph (or matroid) at a particular point of the (x, y)-plane — in other words is a Tutte-Gröthendieck invariant. Other invariants include:

(iv) the chromatic and flow polynomials of a graph;

(v) the partition function of a Q-state Potts model;

(vi) the Jones polynomial of an alternating link;

(vii) the weight enumerator of a linear code over $GF(q)$.

It has been shown that apart from a few special points and 2 special hyperbolae, the exact evaluation of any such invariant is $\#P$-hard even for the very restricted class of planar bipartite graphs. However the question of which points have a fully polynomial randomised approximation scheme is wide open. I shall discuss this problem and give a survey of what is currently known.

The graph terminology used is standard. The complexity theory and notation follows Garey and Johnson (1979). The matroid terminology follows Oxley (1992). Further details of most of the concepts treated here can be found in Welsh (1993).

2. Tutte-Gröthendieck invariants. First consider the following recursive definition of the function $T(G; x, y)$ of a graph G, and two independent variables x, y.

If G has no edges then $T(G; x, y) = 1$, otherwise for any $e \in E(G)$;

(2.1) $T(G; x, y) = T(G'_e; x, y) + T(G''_e; x, y)$, where G'_e denotes the deletion of the edge e from G and G''_e denotes the contraction of e in G,

(2.2) $T(G; x, y) = xT(G'_e; x, y)$ e an isthmus,

(2.3) $T(G; x, y) = yT(G''_e; x, y)$ e a loop.

From this, it is easy to show by induction that T is a 2-variable polynomial in x, y, which we call the *Tutte polynomial* of G.

In other words, T may be calculated recursively by choosing the edges in *any* order and repeatedly using (2.1-3) to evaluate T. The remarkable

---

* Merton College and the Mathematical Institute, Oxford University, Oxford OX1 4JD, England.

fact is that T is well defined in the sense that the resulting polynomial is independent of the order in which the edges are chosen.

Alternatively, and this is often the easiest way to prove properties of T, we can show that T has the following expansion.

First recall that if $A \subseteq E(G)$, the *rank* of A, $r(A)$ is defined by

$$(2.4) \qquad r(A) = |V(G)| - k(A),$$

where $k(A)$ is the number of connected components of the graph $G : A$ having vertex set $V = V(G)$ and edge set A.

It is now straightforward to prove:

(2.5) The Tutte polynomial $T(G; x, y)$ can be expressed in the form

$$T(G; x, y) = \sum_{A \subseteq E} (x - 1)^{r(E) - r(A)} (y - 1)^{|A| - r(A)}.$$

This relates T to the *Whitney rank generating function* $R(G; u, v)$ which is a 2-variable polynomial in the variables u, v, and is defined by

$$(2.6) \qquad R(G; u, v) = \sum_{A \subseteq E} u^{r(E) - r(A)} v^{|A| - r(A)}.$$

It is easy and useful to extend these ideas to matroids.

A *matroid* M is just a pair (E, r) where E is a finite set and r is a submodular *rank function* mapping $2^E \to \mathbf{Z}$ and satisfying the conditions

$$(2.7) \qquad 0 \leq r(A) \leq |A| \qquad A \subseteq E,$$

$$(2.8) \qquad A \subseteq B \Rightarrow r(A) \leq r(B),$$

$$(2.9) \qquad r(A \cup B) + r(A \cap B) \leq r(A) + r(B) \qquad A, B \subseteq E.$$

The edge set of any graph G with its associated rank function as defined by (2.4) is a matroid, but this is just a very small subclass of matroids:- known as graphic matroids.

Given $M = (E, r)$ the *dual matroid* $M^* = (E, r^*)$ where r^* is defined by

$$(2.10) \qquad r^*(E \backslash A) = |E| - r(E) - |A| + r(A).$$

We now just extend the definition of the Tutte polynomial from graphs to matroids by,

$$(2.11) \qquad T(M;x,y) = \sum_{A \subseteq E(M)} (x-1)^{r(E)-r(A)} (y-1)^{|A|-r(A)}.$$

Much of the theory developed for graphs goes through in this more general setting and there are many other applications as we shall see. For example, routine checking shows that

$$(2.12) \qquad T(M;x,y) = T(M^*;y,x).$$

In particular, when G is a planar graph and G^* is any plane dual of G, (2.12) becomes

$$(2.13) \qquad T(G;x,y) = T(G^*;y,x).$$

A set X is *independent* if $r(X) = |X|$, it is a *base* if it is a maximal independent subset of E. An easy way to work with the dual matroid M^* is not via the rank function but by the following definition.

(2.14) M^* has as its bases all sets of the form $E \backslash B$, where B is a base of M.

We close this section with what I call the "recipe theorem" from Oxley and Welsh (1979). Its crude interpretation is that whenever a function f on some class of matroids can be shown to satisfy an equation of the form $f(M) = af(M'_e) + b(M''_e)$ for some $e \in E(M)$, then f is essentially an evaluation of the Tutte polynomial. More precisely it says:

(2.15) THEOREM. Let $\mathcal{C}$ be a class of matroids which is closed under direct sums and the taking of minors and suppose that f is well defined on $\mathcal{C}$ and satisfies

$$(2.16) \qquad f(M) = af(M'_e) + bf(M''_e) \qquad e \in E(M)$$
$$(2.17) \qquad f(M_1 \oplus M_2) = f(M_1)f(M_2)$$

then f is given by

$$f(M) = a^{|E|-r(E)} b^{r(E)} T(M; \frac{x_0}{b}, \frac{y_0}{a})$$

where x_0 and y_0 are the values f takes on coloops and loops respectively.

Any invariant f which satisfies (2.16)-(2.17) is called, a *Tutte-Gröthendieck (TG)-invariant*.

Thus, what we are saying is that any TG-invariant has an interpretation as an evaluation of the Tutte polynomial.

3. Reliability and flows. We illustrate the above with two applications.

Reliability theory deals with the probability of points of a network being connected when individual links or edges are unreliable. Early work in the area was by Moore and Shannon (1956) and now it has a huge literature, see for example Colbourne (1987).

Let G be a connected graph in which each edge is independently *open* with probability p and *closed* with probability $q = 1 - p$. The (*all terminal*) *reliability* $R(G; p)$ denotes the probability that in this random model there is a path between each pair of vertices of G. Thus

$$(3.1) \qquad R(G; p) = \sum_A p^{|A|}(1 - p)^{|E \setminus A|}$$

where the sum is over all subsets A of edges which contain a spanning tree of G, and $E = E(G)$.

It is immediate from this that R is a polynomial in p and a simple conditioning argument shows the following connection with the Tutte polynomial.

(3.2)   If G is a connected graph and e is not a loop or coloop then

$$R(G; p) = qR(G'_e; p) + pR(G''_e; p).$$

where $q = 1 - p$.

Using this with the recipe Theorem 2.16 it is straightforward to check the following statement.

(3.3)   Provided G is a connected graph,

$$R(G; p) = q^{|E| - |V| + 1} p^{|V| - 1} T(G; 1, q^{-1}).$$

We now turn to flows. Take any graph G and orient its edges arbitrarily. Take any finite Abelian group H and call a mapping $\phi : E(G) \to H \setminus \{0\}$ a *flow* (or an *H-flow*) if Kirchhoff's laws are obeyed at each vertex of G, the algebra of course being that of the group H.

Note: Standard usage is to describe what we call an H-flow a *nowhere zero* H-flow.

The following statement is somewhat surprising.

(3.4)   The number of H-flows on G depends only on the order of H and not on its structure.

This is an immediate consequence of the fact that the number of flows is a TG-invariant. To see this, let $F(G; H)$ denote the number of H-flows on G. Then a straightforward counting argument shows that the following is true.

(3.5)   Provided the edge e is not an isthmus or a loop of G then

$$F(G; H) = F(G_e''; H) - F(G_e'; H).$$

Now it is easy to see that if C, L represents respectively a coloop ($=$ isthmus) and loop then

(3.6) $F(C; H) = 0 \quad F(L; H) = o(H) - 1$

where $o(H)$ is the order of H.

Accordingly we can apply the recipe theorem and obtain:

(3.7)   For any graph G and any finite abelian group H,

$$F(G; H) = (-1)^{|E|-|V|+k(G)} T(G; 0, 1 - o(H)).$$

The observation (3.4) is an obvious corollary.

A consequence of this is that we can now speak of G *having a k-flow* to mean that G has a flow over *any* or equivalently *some* Abelian group of order k.

Moreover it follows that there exists a polynomial $F(G; \lambda)$ such that if H is Abelian of order k, then $F(G; H) = F(G; k)$. We call F the *flow polynomial* of G.

The duality relationship (2.12) gives:

(3.8) If G is planar then the flow polynomial of G is essentially the chromatic polynomial of G^*, in the sense that

$$\lambda^{k(G)} F(G; \lambda) = P(G^*; \lambda).$$

A consequence of this and the Four Colour Theorem is that:

(3.9) Every planar graph having no isthmus has a 4-flow.

What is much more surprising is that the following statement is believed to be true:

(3.10)   **Tutte's 5-Flow Conjecture**: Any graph having no isthmus has a 5-flow.

It is far from obvious that there is any universal constant k such that graphs without isthmuses have a k-flow. However Seymour (1981) showed:

(3.11) THEOREM. Every graph having no isthmus has a 6-flow.

For more on this and a host of related graph theoretic problems we refer to Jaeger (1988a).

4. A catalogue of invariants. We now collect together some of the naturally occurring interpretations of the Tutte polynomial. Throughout G is a graph, M is a matroid and E will denote $E(G), E(M)$ respectively.

(4.1)   At $(1,1)$ T counts the number of bases of M (spanning trees in a connected graph).

(4.2)   At $(2,1)$ T counts the number of independent sets of M, (forests in a graph).

(4.3)   At (1,2) T counts the number of spanning sets of M, that is
        sets which contain a base.

(4.4)   At (2,0), T counts the number of acyclic orientations of G.
        Stanley (1973) also gives interpretations of T at $(m, 0)$ for gen-
        eral positive integer m, in terms of acyclic orientations.

(4.5)   Another interpretation at (2,0), and this for a different class
        of matroids, was discovered by Zaslavsky (1975). This is in
        terms of counting the number of different arrangements of sets
        of hyperplanes in n-dimensional Euclidean space.

(4.6) $T(G; -1, -1) = (-1)^{|E|}(-2)^{d(B)}$ where B is the bicycle space
        of G, see Read and Rosenstiehl (1978). When G is planar it
        also has interpretations in terms of the Arf invariant of the
        associated knot.

(4.7)   The chromatic polynomial $P(G; \lambda)$ is given by

$$P(G; \lambda) = (-1)^{r(E)} \lambda^{k(G)} T(G; 1 - \lambda, 0)$$

        where $k(G)$ is the number of connected components.

(4.8)   The flow polynomial $F(G; \lambda)$ is given by

$$F(G; \lambda) = (-1)^{|E|-r(E)} T(G; 0, 1 - \lambda).$$

(4.9)   The (all terminal) reliability $R(G : p)$ is given by

$$R(G; p) = q^{|E|-r(E)} p^{r(E)} T(G; 1, 1/q)$$

        where $q = 1 - p$.

        In each of the above cases, the interesting quantity (on the left hand
side) is given (up to an easily determined term) by an evaluation of the
Tutte polynomial. We shall use the phrase "*specialises to*" to indicate this.
Thus for example, along $y = 0$, T specialises to the chromatic polynomial.

        It turns out that the hyperbolae H_α defined by

$$H_\alpha = \{(x, y) : (x - 1)(y - 1) = \alpha\}$$

seem to have a special role in the theory. We note several important spe-
cialisations below.

(4.10) Along H_1, $T(G; x, y) = x^{|E|}(x - 1)^{r(E)-|E|}$.

(4.11) Along H_2; when G is a graph T specialises to the partition
        function of the Ising model.

(4.12) Along H_q, for general positive integer q, T specialises to the
        partition function of the Potts model of statistical physics.

(4.13) Along H_q, when q is a prime power, for a matroid M of vec-
        tors over $GF(q)$, T specialises to the weight enumerator of the
        linear code over $GF(q)$, determined by M. Equation (2.12)
        relating $T(M)$ to $T(M^*)$ gives the MacWilliams identity of
        coding theory.

(4.14) Along H_q for any positive, not necessarily integer, q, T specialises to the partition function of the random cluster model introduced by Fortuin and Kasteleyn (1971).

(4.15) Along the hyperbola $xy = 1$ when G is planar, T specialises to the Jones polynomial of the alternating link or knot associated with G. This connection was first discovered by Thistlethwaite (1987).

Other more specialised interpretations can be found in the survey of Brylawski and Oxley (1992).

5. The complexity of the Tutte plane. We have seen that along different curves of the x, y plane, the Tutte polynomial evaluates such diverse quantities as reliability probabilities, the weight enumerator of a linear code, the partition function of the Ising and Potts models of statistical physics, the chromatic and flow polynomials of a graph, and the Jones polynomial of an alternating knot. Since it is also the case that for particular curves and at particular points the computational complexity of the evaluation can vary from being polynomial time computable to being $\#P$-hard a more detailed analysis of the complexity of evaluation is needed in order to give a better understanding of what is and is not computationally feasible for these sort of problems. The section is based on the paper of Jaeger, Vertigan and Welsh (1990) which will henceforth be referred to as [JVW].

First consider the problem:

$\pi_1[\mathcal{C}]$: *TUTTE POLYNOMIAL OF CLASS $\mathcal{C}$*

INSTANCE: Graph G belonging to the class $\mathcal{C}$.

OUTPUT: The coefficients of the Tutte polynomial of G.

We note first that for all but the most restricted classes this problem will be $\#P$-hard. This follows from the following observations.

(5.1) Determining the Tutte polynomial of a planar graph is $\#P$-hard.

Proof. Determining the chromatic polynomial of a planar graph is $\#P$-hard and this problem is the evaluation of the Tutte polynomial along the line $y = 0$. $\square$

It follows that:

(5.2) If $\mathcal{C}$ is any class of graphs which contains all planar graphs then $\pi_1[\mathcal{C}]$ is $\#P$-hard.

However it does not follow that it may not be easy to determine the value of $T(G; x, y)$ at particular points or along particular curves of the x, y plane. For example, the evaluation of the Tutte polynomial at (1,1) gives the number of spanning trees of the underlying graph and hence the Kirchhoff determinantal formula shows:

(5.3)   Evaluating $T(G; 1, 1)$ for general graphs is in P.

We now consider two further problems.

$\pi_2[\mathcal{C} : L]$ *TUTTE POLYNOMIAL OF CLASS $\mathcal{C}$ ALONG CURVE L*

INSTANCE: Graph G belonging to the class $\mathcal{C}$.

OUTPUT: The Tutte polynomial along the curve L as a polynomial with rational coefficients.

$\pi_3[\mathcal{C} : a, b]$ *TUTTE POLYNOMIAL OF CLASS $\mathcal{C}$ AT (a, b)*

INSTANCE: Graph G belonging to the class $\mathcal{C}$.

OUTPUT: Evaluation of $T(G; a, b)$.

Note: There are some technical difficulties here, inasmuch as we have to place some restriction on the sort of numbers on which we do our arithmetic operations, and also the possible length of inputs. Thus we restrict our arithmetic to be within a field F which is a finite dimensional algebraic extension of the rationals. We also demand (for reasons which become apparent) that F contains the complex numbers i and $e^{2\pi i/3}$. Similarly we demand that any curve L under discussion will be a rational algebraic curve over such a field F and that L is given in standard parametric form. For more details see [JVW] and for more on this general question see Grötschel, Lovász and Schrijver (1988).

Now it is obvious that for any class $\mathcal{C}$, if evaluating T at (a, b) is hard and $(a, b) \in L$ then evaluating T along L is hard. Similarly if evaluating T along L is hard then determining T is hard. In other words:

(5.4)   For any class $\mathcal{C}$, curve L and point (a, b) with $(a, b) \in L$,

$$\pi_3[\mathcal{C}; a, b] \propto \pi_2[\mathcal{C}; L] \propto \pi_1[\mathcal{C}].$$

Two of the main results of [JVW] are that except when (a, b) is one of a few very special points and L one of a special class of hyperbolae, then the reverse implications hold in (5.4).

Before we can state the two main theorems from [JVW] we need one more definition. Call a class of graphs *closed* if it is closed under the operations of taking minors and series and parallel extensions. That is $\mathcal{C}$ shall remain closed under the four operations of deletion and contraction of an edge together with the insertion of an edge in series or in parallel with an existing edge.

The first result of [JVW] relates evaluations in general with evaluation along a curve.

(5.5)   THEOREM. If $\mathcal{C}$ is any closed class then the problem $\pi_1[\mathcal{C}]$ of determining the Tutte polynomial of members of $\mathcal{C}$ is polynomial time reducible to the problem $\pi_2[C; L]$ for any curve L, except when L is one of the hyperbolae defined by

$$H_\alpha \equiv (x-1)(y-1) = \alpha \qquad \alpha \neq 0,$$

or the degenerate hyperbolae

$$H_0^x \equiv \{(x,y) : x = 1\}$$
$$H_0^y \equiv \{(x,y) : y = 1\}.$$

We call the hyperbolae H_α special hyperbolae and an immediate corollary of the theorem is:

(5.6) COROLLARY. If L is a curve in the x, y plane which is not one of the special hyperbolae, and $\pi_1(C)$ is #P-hard then $\pi_1(C; L)$ is #P-hard.

The second main theorem of [JVW] relates the complexity of determining the Tutte polynomial along a curve with determining its value at a particular point on the curve.

(5.7) THEOREM. The problem $\pi_3[C; a, b]$ of evaluating $T(G; a, b)$ for members G of C (a closed class) is polynomial time reducible to evaluating T along the special hyperbola through (a, b) except when (a, b) is one of the special points $(1,1),(0,0),(-1,-1),(0,-1),(-1,0),(i,-i),(-i,i)$ and $(j, j^2), (j^2, j)$ where $i^2 = -1$ and $j = e^{2\pi i/3}$.

In other words unless a point is one of these 9 special points evaluating the Tutte polynomial at that point is no easier than evaluating it along the special hyperbola through that point.

As an illustration of the applicability of these results we state without proof the following theorem from [JVW].

(5.8) THEOREM. The problem of evaluating the Tutte polynomial of a graph at a point (a, b) is #P-hard except when (a, b) is on the special hyperbola

$$H_1 \equiv (x-1)(y-1) = 1$$

or when (a, b) is one of the special points $(1,1),(-1,-1),(0,-1),(-1,0),(i,-i),(-i,i),(j,j^2)$ and (j^2,j), where $j = e^{2\pi i/3}$. In each of these exceptional cases the evaluation can be done in polynomial time.

As far as planar graphs are concerned, there is a significant difference. The technique developed using the Pfaffian to solve the Ising problem for the plane square lattice by Fisher (1966) and Kasteleyn (1961) can be extended to give a polynomial time algorithm for the evaluation of the Tutte polynomial of any planar graph along the special hyperbola

$$H_2 \equiv (x - 1)(y - 1) = 2.$$

Thus this hyperbola is also "easy" for planar graphs. However it is easy to see that H_3 cannot be easy for planar graphs since it contains the point (-2,0) which counts the number of 3-colourings and since deciding whether a planar graph is 3-colourable is NP-hard, this must be at least NP-hard. However it does not seem easy to show that H_4 is hard for planar graphs. The decision problem is after all trivial by the four colour theorem. The fact that it is $\#P$-hard is just part of the following extension of Theorem (5.8) due to Vertigan and Welsh (1992)

(5.9) THEOREM. The evaluation of the Tutte polynomial of bipartite planar graphs at a point (a, b) is $\#P$-hard except when

$$(a, b) \in H_1 \cup H_2 \cup \{(1, 1), (-1, -1), (j, j^2), (j^2, j)\}$$

when it is computable in polynomial time.

6. Approximating to within a ratio.
We know that computing the number of 3-colourings of a graph G is $\#P$- hard. It is natural therefore to ask how well can we approximate it?

For positive numbers a and $r \geq 1$, we say that a third quantity $\hat{a}$ *approximates a within ratio r* or is an *r-approximation* to a, if

(6.1) $$r^{-1}a \leq \hat{a} \leq ra.$$

In other words the ratio $\hat{a}/a$ lies in $[r^{-1}, r]$.

Now consider what it would mean to be able to find a polynomial time algorithm which gave an approximation within r to the number of 3-colourings of a graph. We would clearly have a polynomial time algorithm which would decide whether or not a graph is 3-colourable. But this is NP-hard. Thus no such algorithm can exist unless $NP = P$.

But we have just used 3-colouring as a typical example and the same argument can be applied to any function which counts objects whose existence is NP- hard to decide. In other words:

(6.2) PROPOSITION. If $f : \Sigma^* \to N$ is such that it is NP-hard to decide whether $f(x)$ is non-zero, then for any constant r there cannot exist a polynomial time r-approximation to f unless $NP = P$.

We now turn to consider a randomised approach to counting problems and make the following definition.

An *ε-δ-approximation scheme* for a counting problem f is a Monte Carlo algorithm which on every input $\langle x, \epsilon, \delta \rangle$, $\epsilon > 0$, $\delta > 0$, outputs a number $\tilde{Y}$ such that

$$Pr\{(1 - \epsilon)f(x) \leq \tilde{Y} \leq (1 + \epsilon)f(x)\} \geq 1 - \delta.$$

It is important to emphasize that there is no mention of running time in this definition.

Now let f be a function from input strings to the natural numbers. A *randomised approximation scheme* for f is a probabilistic algorithm that takes as an input a string x and a rational number ϵ, $0 < \epsilon < 1$, and produces as output a random variable Y, such that Y approximates $f(x)$ within ratio $1 + \epsilon$ with probability $\geq 3/4$.

In other words,

$$(6.3) \qquad Pr\left\{ \frac{1}{1 + \epsilon} \leq \frac{Y}{f(x)} \leq 1 + \epsilon \right\} \geq \frac{3}{4}.$$

A *fully polynomial randomised approximation scheme* (fpras) for a function $f : \Sigma^* \to N$ is a randomised approximation scheme which runs in time which is a polynomial function of n and ϵ^{-1}.

Suppose now we have such an approximation scheme and suppose further that it works in polynomial time. Then we can boost the success probability up to $1 - \delta$ for any desired $\delta > 0$, by using the following trick of Jerrum, Valiant and Vazirani (1986). This consists of running the algorithm $O(\log \delta^{-1})$ times and taking the median of the results.

We make this precise as follows:

(6.4) PROPOSITION. If there exists a fpras for computing f then there exists an $\epsilon - \delta$ approximation scheme for f which on input $\langle x, \epsilon, \delta \rangle$ runs in time which is bounded by $O(\log \delta^{-1})\text{poly}(x, \epsilon^{-1})$.

It is worth emphasising here that the existence of a fpras for a counting problem is a very strong result, it is the analogue of an RP algorithm for a decision problem and corresponds to the notion of tractability. However we should also note, that by an analogous argument to that used in proving Proposition (6.2) we have:

(6.5) PROPOSITION. If $f : \Sigma^* \to N$ is such that deciding if f is nonzero is NP-hard then there cannot exist a fpras for f unless NP is equal to random polynomial time RP.

Since this is thought to be most unlikely, it makes sense only to seek out a fpras when counting objects for which the decision problem is not NP-hard.

Hence we have immediately from the NP-hardness of k-colouring that:

(6.6)   Unless $NP = RP$ there cannot exist a fpras for evaluating $T(G; -k, 0)$ for any integer $k \geq 2$.

Recall now from (4.12) that along the hyperbola, H_Q, for positive integer Q, T evaluates the partition function of the Q-state Potts model.

In an important paper, Jerrum and Sinclair (1990), have shown that there exists a fpras for the ferromagnetic Ising problem. This corresponds to the $Q = 2$ Potts model and thus, their result can be restated in the terminology of this paper as follows.

(6.7)   There exists a fpras for estimating T along the positive branch of the hyperbola H_2.

However it seems to be difficult to extend the argument to prove a similar result for the Q-state Potts model with $Q > 2$ and this remains one of the outstanding open problems in this area.

A second result of Jerrum and Sinclair is the following:

(6.8)   There is no fpras for estimating the antiferromagnetic Ising partition function unless $NP = RP$.

Since it is regarded as highly unlikely that $NP = RP$, this can be taken as evidence of the intractability of the antiferromagnetic problem.

Examination of (6.8) in the context of its Tutte plane representation shows that it can be restated as follows.

(6.9)   Unless $NP = RP$, there is no fpras for estimating T along the curve

$$\{(x, y) : (x - 1)(y - 1) = 2, \quad 0 < y < 1\}.$$

The following extension of this result is proved in Welsh (1993b).

(6.10) THEOREM. On the assumption that $NP \neq RP$, the following statements are true.

(6.11)   Even in the planar case, there is no fully polynomial randomised approximation scheme for T along the negative branch of the hyperbola H_3.

(6.12)   For $Q = 2, 4, 5, ...$, there is no fully polynomial randomised approximation scheme for T along the curves

$$H_Q^- \cap \{x < 0\}.$$

It is worth emphasising that the above statements do not rule out the possibility of there being a fpras at *specific points* along the negative hyperbolae. For example;

(6.13) T can be evaluated exactly at $(-1, 0)$ and $(0, -1)$ which both lie on H_2^-.

(6.14)   There is no inherent obstacle to there being a fpras for estimating the number of 4-colourings of a planar graph.

I do not believe such a scheme exists but cannot see how to prove it. It certainly is not ruled out by any of our results. I therefore pose the specific question:

(6.15) **Problem.** Is there a fully polynomial randomised approximation scheme for counting the number of k-colourings of a planar graph for any fixed $k \geq 4$?

I conjecture that the answer to (6.15) is negative.

Similarly, since by Seymour's theorem, every bridgeless graph has a nowhere zero 6-flow, there is no obvious obstacle to the existence of a fpras for estimating the number of k-flows for $k \geq 6$. Thus a natural question, which is in the same spirit is the following.

(6.16) Show that there does not exist a fpras for estimating T at $(0, -5)$. More generally, show that there is no fpras for estimating the number of k-flows for $k \geq 6$.

Again, although because of Theorem 6.10, a large section of the relevant hyperbola has no fpras, there is nothing to stop such a scheme existing at isolated points.

Another point of special interest is $(0, -2)$. Mihail and Winkler (1991) have shown, among other things, that there exists a fpras for counting the number of ice configurations in a 4-regular graph. This is equivalent to the statement:

(6.17) For 4-regular graphs counting nowhere-zero 3-flows has a fpras.

In other words:

(6.18) There is a fpras for computing T at $(0, -2)$ for 4-regular graphs.

The reader will note that all these 'negative results' are about evaluations of T in the region outside the quadrant $x \geq 1$, $y \geq 1$. In Welsh (1993) it is conjectured that the following is true:

(6.19) *Conjecture.* There exists a fpras for evaluating T at all points of the quadrant $x \geq 1$, $y \geq 1$.

Some further recent evidence in support of this is given in the next section.

7. Denseness helps. One mild curiosity about quantities which are hard to count is that the counting problem seems to be easier when the underlying structure is dense. A well known example of this is in approximating the permanent. Less widely known is the result of Edwards (1986) which shows that it is possible to exactly count the number of k-colourings of graphs in which each vertex has at least $(k - 2)/(k - 1)$ neighbours.

More precisely, if we let $\mathcal{G}_\alpha$ be the collection of graphs $G = (V, E)$ such that each vertex has at least $\alpha |V|$ neighbours then we say G is *dense* if $G \in \mathcal{G}_\alpha$ for some α. Edwards shows:

(7.1) For $G \in \mathcal{G}_\alpha$, evaluation of $T(G; 1 - k, 0)$ is in P provided $\alpha > (k - 2)/(k - 1)$.

As far as approximation is concerned, a major advance is the recent result of Annan (1993) who showed that:

(7.2) There exists a fpras for counting forests in dense graphs.

Now the number of forests is just the evaluation of T at the point (2.1) and a more general version of (7.2) is the following result, also by Annan.

(7.3) For dense G, there is a fpras for evaluating $T(G; x, 1)$ for positive integer x.

The natural question suggested by (7.3) is about the matroidal dual - namely, does there exist a fpras for evaluating T at $(1, x)$? This is the reliability question, and in particular, the point (1,2) enumerates the number of connected subgraphs. It is impossible to combine duality with denseness so Annan's methods don't seem to work.

It turns out that I can obtain the following result:

(7.4) For dense G there exists a fpras for evaluating $T(G; x, y)$ for all (x, y) in the region

$$\{1 \leq x\} \cap \{1 < y\} \cap \{(x - 1)(y - 1) \leq 1\}.$$

An immediate corollary is:

(7.5) There exists a fpras for estimating the reliability probability in dense graphs.

The proof of (7.4) hinges on a result of Joel Spencer (1993). In its simplest form it can be stated as follows:

(7.6) For $G \in \mathcal{G}_\alpha$, if G_p denotes a random subgraph of G obtained by deleting edges from G with probability $1 - p$, independently for each edge, then there exists $d(\alpha, p) < 1$, such that

$$Pr\{G_p \text{ is disconnected}\} \leq d(\alpha, p).$$

At this meeting, I have discovered that Alan Frieze has also obtained a result which is very similar to (7.5), and since the meeting, using some ideas of Noga Alon, together we have been able to eliminate the constraint $(x - 1)(y - 1) \leq 1$ in (7.4). We aim to write this up in the near future.

I close with three open problems related to (7.4).

(7.7) Does there exist a fpras for counting (a) acyclic orientations, (b) forests of size k, or (c) connected subgraphs of size k, in dense graphs?

Two of these problems are particular instances of the problem of estimating the number of bases in a matroid of some class. For example the question about forests is just counting the number of bases in truncations of graphic matroids.

A very interesting new light on such problems is outlined in the recent paper of Feder and Mihail (1992) who show that a sufficient condition for the natural random walk on the set of bases of M to be rapidly mixing is that M and its minors are *negatively correlated*. That is, they satisfy the constraint, that if B_R denotes a random base of M and e, f is any pair of distinct elements of $E(M)$, then

$$Pr\{e \in B_R \mid f \in B_R\} \leq P_r\{e \in B_R\}.$$

This concept of negative correlation goes back to Seymour and Welsh (1975). This was concerned with the Tutte polynomial and the log-concave nature of various of its evaluations and coefficients. It was shown there that not all matroids have this property of negative correlation; however graphic and cographic ones do, and it is mildly intrigueing that this concept has reappeared now in an entirely new context.

Acknowledgement. The original impetus to look at evaluations of T in the dense case, arose in discussions with Marek Karpinski in a RAND meeting in Oxford in September 1992. Further discussions with Joel Spencer, Alan Frieze, Noga Alon, Nabil Kahale and James Annan have also been very helpful.

REFERENCES

1. ANNAN, J.D., *A randomised approximation algorithm for counting the number of forests in dense graphs*, Combinatorics, Probability and Computer Science, (submitted) (1993).

2. BRYLAWSKI, T.H. AND OXLEY, J.G., *The Tutte polynomial and its applications*, Matroid Applications (ed. N. White), Cambridge Univ. Press (1992), pp. 123–225.

3. COLBOURN, C.J., *The Combinatorics of Network Reliability*, Oxford University Press (1987).

4. EDWARDS, K., *The complexity of colouring problems on dense graphs*, Theoretical Computer Science, **43** (1986), pp. 337–343.

5. FEDER, T. AND MIHAIL, M., *Balanced matroids*, Proceedings of 24th Annual ACM Symposium on the Theory of Computing (1992), pp. 26–38.

6. GAREY, M.R. AND JOHNSON, D.S., *Computers and Intractability - A guide to the theory of NP-completeness.* Freeman, San Francisco (1979).

7. GRÖTSCHEL, M., LOVÁSZ, L. AND SCHRIJVER, A., *Geometric Algorithms and Combinatorial Optimization*, Springer-Verlag, Berlin (1988).

8. JAEGER, F., *Nowhere zero flow problems*, Selected Topics in Graph Theory 3, (ed. L. Beineke and R.J. Wilson) Academic Press, London (1988), pp. 71–92.

9. JAEGER, F., VERTIGAN, D.L., AND WELSH, D.J.A., *On the computational complexity of the Jones and Tutte polynomials*, Math. Proc. Camb. Phil. Soc. **108** (1990), pp. 35–53.

10. JERRUM, M.R. AND SINCLAIR, A., *Polynomial-time approximation algorithms for the Ising model*, Proc. 17th ICALP, EATCS (1990), pp. 462–475.

11. JERRUM, M.R., VALIANT, L.G. AND VAZIRANI, V.V., *Random generation of combinatorial structures from a uniform distribution*, Theoretical Computer Science, **43** (1990), pp. 169–188.

12. MIHAIL, M. AND WINKLER, P., *On the number of Eulerian orientations of a graph*, Bellcore Technical Memorandum TM-ARH-018829 (1991).

13. MOORE, E.F. AND SHANNON, C.E., *Reliable circuits using less reliable components*, Journ. Franklin Instit., **262** (1956), pp. 191–208, 281–297.

14. OXLEY, J.G., *Matroid Theory*, Oxford Univ. Press (1992).

15. OXLEY, J.G. AND WELSH, D.J.A., (1979) *The Tutte polynomial and percolation*, Graph Theory and Related Topics (eds. J.A. Bondy and U.S.R. Murty), Academic Press, London (1979), pp. 329–339.

16. ROSENSTIEHL, P. AND READ, R.C., *On the principal edge tripartition of a graph*, Ann. Discrete Math., 3 (1978), pp. 195–226.

17. SEYMOUR, P.D., *Nowhere-zero 6-flows*, J. Comb. Theory **B**, **30**, (1981), pp. 130–135.

18. SEYMOUR, P.D. AND WELSH, D.J.A., *Combinatorial applications of an inequality from statistical mechanics*, Math. Proc. Camb. Phil. Soc., **77** (1975), pp. 485–497.

19. SPENCER, J. (private communication) (1993).

20. THISTLETHWAITE, M.B., *A spanning tree expansion of the Jones polynomial*, Topology, **26** (1987), pp. 297–309.

21. VERTIGAN, D.L. AND WELSH, D.J.A., *The computational complexity of the Tutte plane: the bipartite case*, Probability, Combinatorics and Computer Science, 1 (1992), pp. 181–187.

22. WELSH, D.J.A., *Complexity: Knots, Colourings and Counting*, London Mathematical Society Lecture Note Series **186**, Cambridge University Press (1993).

23. WELSH, D.J.A., *Randomised approximation in the Tutte plane*, Combinatorics, Probability and Computing, 3 (1994), pp. 137–143.

24. ZASLAVSKY, T., *Facing up to arrangements: face count formulas for partitions of spaces by hyperplanes*, Memoirs of American Mathematical Society, **154** (1975).

QUASI-ADDITIVE EUCLIDEAN FUNCTIONALS[*]

J.E. YUKICH[†]

Abstract. Euclidean functionals having a certain "quasi-additivity" property are shown to provide a general approach to the limit theory of a broad class of random processes which arise in stochastic matching problems. Via the theory of quasi-additive functionals, we obtain a Beardwood-Halton-Hammersley type of limit theorem for the TSP, MST, minimal matching, Steiner tree, and Euclidean semi-matching functionals. One minor but technically useful consequence of the theory is that it often shows that the asymptotic behavior of functionals on the d-dimensional cube coincides with the behavior of the same functional defined on the d-dimensional torus. A more pointed consequence of the theory is that it leads in a natural way to a result on rates of convergence. Finally, we show that the quasi-additive functionals may be approximated by a heuristic with polynomial mean execution time.

1. Introduction. In 1959 Beardwood, Halton, and Hammersley proved a remarkable theorem concerning the stochastic version of the traveling salesman problem (TSP):

THEOREM 1.1. *(Beardwood, Halton, and Hammersley (1959)) If X_i, $i \geq 1$, are i.i.d. random variables with bounded support in R^d, $d \geq 2$, then the length L_n under the usual Euclidean metric of the shortest path through the points $\{X_1, \ldots, X_n\}$ satisfies*

$$(1.1) \qquad \lim_{n \to \infty} L_n / n^{(d-1)/d} = \beta \int f(x)^{(d-1)/d} dx \ \ a.s.,$$

where β is a constant depending only on d and where f is the density of the absolutely continuous part of the distribution of X_i.

Much later Steele (1981) observed that one can abstract from the TSP a few of its basic properties and show that the BHH theorem holds for a wide class of Euclidean functionals. More precisely, let L be a real-valued function defined on the finite subsets $\{x_1, \ldots, x_n\}$ of R^d. If L satisfies homogeneity, translation invariance, monotonicity, and subadditivity then it is called a *subadditive Euclidean functional* (Steele (1981)). Examples include the TSP and Steiner tree functionals.

THEOREM 1.2. *(Steele (1981)) Let L be a subadditive Euclidean functional and U_i, $i \geq 1$, i.i.d. random variables with the uniform distribution on $[0,1]^d$, $d \geq 2$. Then*

$$(1.2) \qquad \lim_{n \to \infty} L(U_1, \ldots, U_n) / n^{(d-1)/d} = C(L, d) \ \ a.s.,$$

where $C(L, d)$ is a constant depending on L and d.

Unfortunately, several well-known functionals (including the minimal spanning tree (MST) and greedy matching functionals) are not monotone

---

[*] Research supported in part by NSF Grant Number DMS-9200656.

[†] Department of Mathematics, Lehigh University, Bethlehem, PA 18015.

and thus not covered by Steele's theorem. However, Rhee (1993) shows that if the monotonicity hypothesis is replaced by a continuity hypothesis

$$(1.3) \quad |L(x_1, \ldots, x_n, x_{n+1}, \ldots, x_{n+k}) - L(x_1, \ldots, x_n)| \leq Ck^{(d-1)/d},$$

then the asymptotic result (1.2) continues to hold and actually holds in the sense of complete convergence. Since continuity (1.3) is easily proved for the MST, TSP, minimal matching and Steiner tree functionals, Rhee extends the scope of Steele's theorem.

This raises an intriguing well-recognized problem: using only *natural assumptions such as continuity, subadditivity, and superadditivity*, extend Steele's theorem to random variables with an *arbitrary* distribution on R^d. The extension should hold for non-monotone functionals, should ideally reveal the intrinsic character of subadditive Euclidean functionals, and thus capture the essence of the BHH theorem.

The work of Redmond and Yukich (1993) on functionals which we term "quasi-additive" essentially solves this problem. The present paper begins with a review of the theory of quasi-additive Euclidean functionals as initiated in Redmond and Yukich (1993). We then show how such functionals provide a general approach to the limit theory of Euclidean functionals, including the TSP, MST, minimal matching, Steiner tree, and Euclidean semi-matching functionals. Moreover, quasi-additive functionals also admit a rate of convergence, and thus add to the rate results of Alexander (1993). In this way quasi-additive functionals unify, extend, and simplify the earlier work of Steele (1981), Rhee (1993), and Alexander (1993).

Quasi-additive functionals serve additional purposes as well. For example, the theory of quasi-additive functionals shows in a natural way that the asymptotics for functionals on the d-cube coincide with asymptotics on the d-dimensional torus and in this way we recover quite simply the results of Jaillet (1993).

Central to the theory of quasi-additive functionals is that it associates to every functional a dual functional which approximates it in a well defined sense. We find that the dual provides an approximation which has polynomial mean execution time. The approximation plays a role analogous to that played by Karp's polynomial time heuristic for the TSP.

The present paper does not provide complete proofs of results contained in Redmond and Yukich (1993), but results presented here for the first time (Proposition 2.1, Theorem 3.3, and Theorem 4.1) will be proved in their entirety.

2. Quasi-additivity. Let L denote a real-valued function defined on the finite subsets of $[0, 1]^d$, $d \geq 2$, with $L(\phi) = 0$. Say that L is subadditive if

(i) (Subadditivity) There exists a constant C_1 with the following property: If $\{Q_i\}_{i=1}^{m^d}$ denotes a partition of $[0, 1]^d$ into m^d subcubes of edge length m^{-1}, and if $q_i \in R^d$ are chosen such that $m(Q_i - q_i) = [0, 1]^d$, then for every finite subset F of $[0,1]^d$,

$$L(F) \leq m^{-1} \sum_{i \leq m^d} L(m[(F \cap Q_i) - q_i]) + C_1 m^{d-1}.$$

Analogously, say that L is superadditive if

(ii) (Superadditivity) For the same conditions as above on Q_i, m, and q_i we have now

$$L(F) \geq m^{-1} \sum_{i \leq m^d} L(m[(F \cap Q_i) - q_i]) - C_2 m^{d-1}.$$

Notice that if L is translation invariant (i.e., $L(x+F) = L(F)$ for all $x \in R^d$ and every finite subset F of R^d) and homogeneous (i.e., $L(\alpha F) = \alpha L(F)$ for every $\alpha > 0$ and every finite subset F of R^d), then the subadditivity condition reduces to the usual one:

$$L(F) \leq \sum_{i \leq m^d} L(F \cap Q_i) + C_1 m^{d-1}.$$

A similar comment applies to the superadditivity condition. One advantage of the present formulation is that it does not require translation invariance and homogeneity of L, and although this distinction may at first seem purely formal, we will see shortly that it offers telling practical consequences.

Finally, say that L is continuous if

(iii) (Continuity) There exists a constant C_3 such that for all finite subsets F and G of $[0,1]^d$,

$$|L(F \cup G) - L(F)| \leq C_3(\text{card } G)^{(d-1)/d}.$$

As a simple consequence of continuity, notice that

$$|L(G)| \leq C_3(\text{card } G)^{(d-1)/d}.$$

It is usually difficult to verify that a functional L simultaneously satisfies subadditivity and superadditivity. We distinguish between these possibilities and agree to say that if L satisfies (i) and (iii), then it is a *continuous subadditive Euclidean functional*; if L satisfies (ii) and (iii), then it is a *continuous superadditive Euclidean functional*.

The central fact which makes the present theory useful is that many continuous subadditive Euclidean functionals L on $[0,1]^d$ are naturally related to a dual superadditive Euclidean functional $\hat{L}$ that is itself a good approximation to L. In particular we find that we can require $\hat{L}(F) \leq 1 + L(F)$ for all F together with the bound

(2.1) $|EL(U_1, \ldots, U_n) - E\hat{L}(U_1, \ldots, U_n)| \leq C_4 n^{(d-2)/d},$

where here and elsewhere $U_1, \ldots, U_n$ denote i.i.d. uniform random variables on $[0,1]^d$. As we shall see shortly, $\hat{L}$ is typically a boundary-rooted version of L, one where edges may be connected to the boundary of the unit cube. Thus $\hat{L}$ is typically neither translation invariant nor homogeneous. When the continuous subadditive Euclidean functional L and its dual $\hat{L}$ enjoy the approximation property (2.1) we say that L (as well as $\hat{L}$) is a "quasi-additive continuous Euclidean functional".

Quasi-additive functionals form a pleasantly broad class.

THEOREM 2.1. *The TSP, MST, Steiner tree, Euclidean minimal matching problem, and Euclidean semi-matching problem all yield quasi-additive functionals.*

Before sketching the proof of Theorem 2.1, we first describe the construction of the dual functionals. This process may be illustrated by considering the dual associated to the MST functional $T := T(x_1, \ldots, x_n)$ giving the weight of the minimal spanning tree of $\{x_1, \ldots, x_n\}$, where the weight assigned to an edge is its length.

Define the dual

$$T_r(F) := \min \sum_i T(F_i \cup \{a_i\})$$

where the minimum ranges over all partitions $(F_i)_{i \geq 1}$ of F and all sequences of points $\{a_i\}_{i \geq 1}$ on the boundary of $[0,1]^d$. Thus, $T_r(F)$ denotes the minimum over all partitions $(F_i)_{i \geq 1}$ and all points $\{a_i\}_{i \geq 1}$ of the sum of the lengths of the trees through disjoint subsets F_i of F, where each tree is rooted to the boundary point a_i. The graph realizing $T_r(F)$ may be interpreted as a collection of small trees connected via the boundary into one large tree, where the connections along the boundary incur no cost.

Thus T_r may be considered as the boundary rooted version of T and duals for the other functionals can be obtained through a similar rooting procedure.

One essential point of the construction is that the resulting T_r is superadditive, and we now show that T and T_r are both quasi-additive, i.e., they satisfy the approximation (2.1). The proof rests upon an auxiliary result from Redmond and Yukich (1993).

LEMMA 2.1. *Let $\{U_i\}_{i \leq n}$ be i.i.d. random variables with the uniform distribution on $[0,1]^d$. Then the expected number of points rooted to the boundary in the optimal rooted minimal spanning tree $T_r(U_1, \ldots, U_n)$ is bounded by $Kn^{(d-1)/d}$, where K depends only on d.*

Equipped with Lemma 2.1, we may now show that the approximation (2.1) holds. Consider a rooted minimal spanning tree T which achieves $T_r(U_1, \ldots, U_n)$. Call the points where T is rooted to the boundary of $[0,1]^d$ "marks". By Lemma 2.1, the expected number of "marks" is bounded by $Kn^{(d-1)/d}$. Letting $L(F)$ denote the length of the shortest cycle through the set F, we see that

(2.2) $T(U_1, \ldots, U_n) \leq T_r(U_1, \ldots, U_n) + L(\{\text{marks}\})$

and

(2.3) $T_r(U_1, \ldots, U_n) \leq T(U_1, \ldots, U_n) + 1/2.$

Observe that since the set of marks lies on a subset of dimension $d - 1$, we have

$$L(\{\text{marks}\}) \leq K(\text{card}\{\text{marks}\})^{(d-2)/(d-1)},$$

for a constant K depending only on d, and therefore by Jensen's inequality

$$\begin{aligned} EL(\{\text{marks}\}) &\leq E(\text{card}\{\text{marks }\})^{(d-2)/(d-1)} \\ &\leq K(n^{(d-1)/d})^{(d-2)/(d-1)} \\ &= Kn^{(d-2)/d}. \end{aligned}$$

Taking expectations in (2.2) and (2.3) yields the approximation (2.1) and thus T and T_r are quasi-additive.

The proofs that the TSP, Euclidean matching and Steiner tree functionals are quasi-additive are very similar and can be found in Redmond and Yukich (1993). The remainder of this section focuses on the proof that Euclidean semi-matchings are quasi-additive. Recall that semi-matchings represent a relaxation of the minimal weight matching problem. More precisely, let $G = (V, E)$ be a graph such that for each edge $e \in E$ there is an associated weight w_e. We seek the solutions to the following linear programming problem:

(2.4)
$$\begin{cases} z = \min_x \sum_{e \in E} x_e w_e \\ \text{subject to} \quad \sum_{e \text{ meets } v} x_e \geq 1 \text{ for all } v \in V \\ \text{and } x_e \geq 0 \text{ for all } e \in E. \end{cases}$$

The Euclidean semi-matching problem seeks the solutions to (2.4) where the vertices V of G are points $v_1, \ldots, v_n$ in R^d and where the weight w_e associated with edge $e := (v_i, v_j)$ is the Euclidean distance $|v_i - v_j|$. As in Steele (1992), we note that since the loading factor x_e can only be 0, 1/2, or 1 in a minimal solution to (2.4), one can easily show that any minimal solution consists of a union of isolated edges with loading 1 and a collection of odd cycles that has all edge loading equal to 1/2. We let $S := S(v_1, \ldots, v_n)$ denote the Euclidean semi-matching functional on the vertices $v_1, \ldots, v_n$; thus S equals the sum of the combined edge lengths of the minimal solution.

PROPOSITION 2.1. *The semi-matching functional is quasi-additive.*

Proof. We define the rooted dual S_r analogously to the rooted MST dual T_r:

$$S_r(F) := \min \sum_i S(F_i \cup \{a_i, b_i\}),$$

where the minimum ranges over all partitions $(F_i)_{i\geq 1}$ of F and all sequences of pairs of points $\{a_i, b_i\}$ lying on the boundary of $[0,1]^d$. One can easily see that S_r is superadditive.

Steele (1992) already observed that S is continuous in the sense of (iii) above and we will verify that the rooted dual S_r is continuous as well. First, observe that for all sets F and G

$$S_r(F \cup G) \leq S_r(F) + S_r(G)$$
$$\leq S_r(F) + S(G) + 1/2$$
$$\leq S_r(F) + C(\text{card} G)^{(d-1)/d}.$$

For the reverse inequality, let F_1 be the set of points of F which are connected to points of G and let $F' := F\backslash F_1$. We see card $F_1 \leq 2\text{card } G$ and by considering the restriction of the optimal semi-matching of $F \cup G$ to F' we find

$$S_r(F) \leq S_r(F') + S_r(F_1)$$
$$\leq S_r(F \cup G) + C(\text{card } F_1)^{(d-1)/d}.$$

On combining the above inequalities, we see

$$|S_r(F \cup G) - S_r(F)| \leq C(\text{card } G)^{(d-1)/d},$$

that is, S_r is continuous.

To complete the proof that S and S_r are quasi-additive, we now only need to show that they satisfy the approximation (2.1). However, this follows from an easy modification of the above argument showing that the MST and its rooted dual satisfy (2.1). First, we observe that Lemma 2.1 carries over to rooted semi-matching functionals: namely, the expected number of points rooted to the boundary in the optimal rooted semi-matching $S_r(U_1, \ldots, U_n)$ is bounded by $Kn^{(d-1)/d}$. Therefore, letting S be the semi-matching functional realizing $S_r(U_1, \ldots, U_n)$ and letting "marks" refer to the points where S is rooted to the boundary of $[0,1]^d$, we obtain $E(\text{card}\{\text{marks}\}) \leq Kn^{(d-1)/d}$. Since

$$S(U_1, \ldots, U_n) \leq S_r(U_1, \ldots, U_n) + L(\{\text{marks}\})$$

and

$$S_r(U_1, \ldots, U_n) \leq S(U_1, \ldots, U_n) + 1/2,$$

it follows, after taking expectations, that S and S_r are quasi-additive.

This completes the proof of Proposition 2.3. □

3. The main properties of quasi-additive Euclidean functionals. In this section the main features of quasi-additive functionals are reviewed. The first result shows that quasi-additive functionals behave remarkably like the TSP:

THEOREM 3.1. *(Redmond and Yukich (1993))* If L is a quasi-additive Euclidean functional and X_i, $i \geq 1$, are i.i.d. random variables with bounded support in R^d, $d \geq 2$, then

$$(3.1) \quad \lim_{n \to \infty} L(X_1, \ldots, X_n)/n^{(d-1)/d} = C(L,d) \int f(x)^{(d-1)/d} dx \quad c.c. \ ,$$

where $C(L,d)$ is a constant depending only on L and d, f is the density of the absolutely continuous part of the distribution of X_i, and c.c. denotes complete convergence.

In view of the history of the developments of the BHH theorem, the proof of (3.1), as given in Redmond and Yukich (1993), is surprisingly simple and straightforward. From the subadditivity of L we can see that (3.1) holds for uniformly distributed random variables; this is observed by Rhee (1993). The passage to the general case is greatly simplified by the assumed subadditivity and superadditivity of L and $\hat{L}$, respectively.

The next result shows that the quasi-additive structure readily yields a rate of convergence in the uniform case.

THEOREM 3.2. *a.* *$(d \geq 3)$* If L is a quasi-additive continuous Euclidean functional then

$$(3.2) \quad |EL(U_1, \ldots, U_n) - C(L,d)n^{(d-1)/d}| = O(n^{(d-2)/d}).$$

b. *$(d = 2)$* Let L be a quasi-additive Euclidean functional which also satisfies a "weak continuity" assumption: there is a constant C_5 such that for all $n \geq 1$

$$|EL(U_1, \ldots, U_n) - EL(U_1, \ldots, U_{n+1})| \leq C_5 n^{-1/2}.$$

Then

$$|EL(U_1, \ldots, U_n) - C(L,d)n^{1/2}| = O(1).$$

Although we will not provide the details of the proof, this theorem is a straightforward consequence of sub- and superadditivity. The key observation is that when $d \geq 3$ the subadditivity of L readily implies

$$EL(U_1, \ldots, U_n) - C(L,d)n^{(d-1)/d} \geq -Kn^{(d-1)/d}$$

whereas the superadditivity of the dual $\hat{L}$ implies

$$E\hat{L}(U_1, \ldots, U_n) - C(L,d)n^{(d-1)/d} \leq Kn^{(d-1)/d}.$$

Combining these estimates with the approximation (2.1) gives (3.2).

The next result shows that the asymptotic behavior of the quasi-additive functionals L described by Theorem 2.1 coincides with their asymptotic behavior on the d-dimensional torus. Thus, to find the unknown asymptotic constant $C(L,d)$ it suffices to assume that L is defined on the torus, where the absence of boundary conditions simplifies the approach. Indeed, Avram and Bertsimas (1992) successfully evaluate $C(L, d)$ when L is the minimal spanning tree on the torus.

The following theorem generalizes results of Jaillet (1993) and furnishes a second example of how the use of dual processes leads to proofs that are short and simple.

THEOREM 3.3. *Let L denote either the TSP, MST, Steiner tree, minimal matching, or semi-matching functional and let L^T denote the associated functional on the d-dimensional torus equipped with the flat metric on the d-cube. Let X_i, $i \geq 1$, be i.i.d. random variables with values in the d-cube. One then has*

$$\lim_{n \to \infty} L^T(X_1, \ldots, X_n)/n^{(d-1)/d} = \lim_{n \to \infty} L(X_1, \ldots, X_n)/n^{(d-1)/d}$$
$$= C(L, d) \int f(x)^{(d-1)/d} dx \ \ c.c. \ ,$$

where f is the density of the absolutely continuous part of the distribution of X_i.

Proof. One can easily check that for all sets F

$$\hat{L}(F) \leq L^T(F) + 1 \leq L(F) + 1.$$

Thus the asymptotic behavior of L^T coincides with the asymptotic behavior of L (and $\hat{L}$). Now apply Theorem 3.1. □

4. The dual as a polynomial time approximation. Since the work initiated by Karp (1977), we know that there are efficient methods for approximating the length $L_n := L_n(U_1, \ldots, U_n)$ of the shortest path through i.i.d. uniformly distributed random variables $U_1, \ldots, U_n$; cf. Karp and Steele (1985) for an exposition. More precisely, the fixed dissection algorithm provides a simple heuristic L_n^F having the property that $L_n^F(U_1, \ldots, U_n)/L(U_1, \ldots, U_n)$ converges completely to 1 and which moreover has polynomial expected execution time.

In this section we show that the general framework of quasi-additive functionals and the special emphasis on the duals $\hat{L}$ allows us to construct an approximation $\hat{L}^A$ with properties similar to Karp's heuristic.

Throughout, let L denote a quasi-additive Euclidean functional and $\hat{L}$ its dual. For a fixed m, consider the usual partition $\{Q_i\}_{i=1}^{m^d}$ of $[0, 1]^d$ into subcubes and define the approximation $\hat{L}^A$ based on the dual by

$$(4.1) \qquad \hat{L}^A(F) := m^{-1} \sum_{i \leq m^d} \hat{L}(m[(F \cap Q_i) - q_i]).$$

For any finite subset $F \subset [0, 1]^d$, the preceding definitions tell us that

$$\hat{L}^A(F) \leq \hat{L}(F) \leq 1 + L(F)$$
$$\leq 1 + m^{-1} \sum_{i \leq m^d} L(m[(F \cap Q_i) - q_i]) + C_1 m^{d-1}.$$

Letting $F := \{U_1, \ldots, U_n\}$, taking expectations, and applying the approximation (2.1) gives

$$E\hat{L}^A(F) \leq 1 + EL(F)$$
$$\leq 1 + m^{-1} \sum_{i \leq m^d} (E\hat{L}(m[(F \cap Q_i) - q_i]) + C_4(n/m^d)^{(d-2)/d}) + C_1 m^{d-1}$$
$$\leq 1 + E\hat{L}^A(F) + (C_1 + C_4)m^{d-1}(n/m^d)^{(d-2)/d}$$
$$\leq 1 + E\hat{L}^A(F) + (C_1 + C_4)mn^{(d-2)/d}.$$

Hence, $\hat{L}^A$ is a good approximation for L in expectation.

PROPOSITION 4.1.   *The approximation $\hat{L}^A$ based on the partition $\{Q_i\}_{i=1}^{m^d}$ satisfies*

$$E\hat{L}^A(U_1, \ldots, U_n) \leq 1 + EL(U_1, \ldots, U_n)$$
$$\leq 1 + E\hat{L}^A(U_1, \ldots, U_n) + (C_1 + C_4)mn^{(d-2)/d}.$$

Moreover, the time required to compute $\hat{L}^A(U_1, \ldots, U_n)$ is bounded by

$$T_n = \sum_{i \leq m^d} f(N_i),$$

where N_i denotes the cardinality of $Q_i \cap \{U_1, \ldots, U_n\}$ and where $f(N)$ denotes a bound on the time needed to compute $\hat{L}(F)$, card $F = N$. If we assume that f exhibits exponential growth, say $f(x) = Ax^B 2^x$ for some constants A and B, as would be the case for the TSP, then since the N_i, $1 \leq i \leq m^d$, are binomial random variables, straightforward calculations show that

$$ET_n \sim 4An(n/m^d)^{B-1} e^{n/m^d};$$

see, for example, Karp and Steele (1985). Choosing $m := (n/\log n)^{1/d}$, implies that the expected execution time for the approximation $\hat{L}^A$ is $O(n^2 \log^{B-1} n)$.

The preceding considerations are summarized in the following theorem.

THEOREM 4.1.   *Let $L(U_1, \ldots, U_n)$ be a quasi-additive Euclidean functional defined on the i.i.d. uniform random variables $U_1, \ldots, U_n$ and let $\hat{L}^A(U_1, \ldots, U_n)$ be the approximation (4.1) based on the dual $\hat{L}$. With $m = (n/\log n)^{1/d}$ in (4.1), the approximation $\hat{L}^A$ closely approximates L:*

$$E\hat{L}^A(U_1, \ldots, U_n) \leq 1 + EL(U_1, \ldots, U_n)$$
$$\leq 1 + E\hat{L}^A(U_1, \ldots, U_n) + (C_1 + C_4)(\log n)^{-1/d} n^{(d-1)/d}.$$

Moreover, if $\hat{L}(F)$ may be computed in time bounded by $A(\text{card } F)^B \cdot 2^{\text{card } F}$, then the expected execution time for $\hat{L}^A(U_1, \ldots, U_n)$ is $O(n^2 \log^{B-1} n)$.

5. Concluding remarks. The quasi-additive structure of Euclidean functionals L provides a natural and simple framework which helps to both unify and extend previous results. In particular, we determine the asymptotic behavior of quasi-additive functionals on the d-cube and the d-torus in the context of complete convergence, we find their rates of convergence, and we also supply polynomial time approximants. The presence of the dual functional $\hat{L}$ is central to the theory and leads to proofs which are short and simple. We see that the dual gives rise to natural inequalities which form the analogs of the already existing inequalities associated with L. These inequalities, which readily lend themselves to asymptotic analysis in a natural way, combine to give straightforward estimates of L from above and below, thus yielding BHH limit theorems for a broad class of random processes arising in geometric probability.

6. Acknowledgements. The author thanks Charles Redmond for valuable conversations leading to the present paper. The author would also like to thank both Michel Talagrand, who pointed out the simple proof of Theorem 3.3, and Dimitris Bertsimas, who suggested that the dual could serve as a polynomial time approximation. Finally, the author thanks Michael Steele for comments leading to an improved exposition.

REFERENCES

[1] K. ALEXANDER, *Rates of convergence of means for distance-minimizing subadditive Euclidean functionals*, Annals of Applied Prob., to appear (1993).

[2] J. AVRAM AND D. BERTSIMAS, *The minimum spanning tree constant in geometrical probability and under the independent model: A unified approach*, Annals of Applied Prob., 2 (1992), pp. 113–130.

[3] J. BEARDWOOD, J.H. HALTON, AND J.M. HAMMERSLEY, *The shortest path through many points*, Proc. Cambridge Philos. Soc., 55 (1959), pp. 299–327.

[4] P. JAILLET, *Cube versus torus models and the Euclidean minimum spanning tree constant*, Annals of Applied Prob., 3 (1993), pp. 582–592.

[5] R.M. KARP, *Probabilistic analysis of partitioning algorithms for the traveling salesman problem in the plane*, Math. Oper. Research, 2 (1977), pp. 209–224.

[6] R.M. KARP AND J.M. STEELE, *Probabilistic analysis of heuristics*, in The Traveling Salesman Problem, ed. E.L. Lawler et al., J. Wiley and Sons (1985), pp. 181–205.

[7] C. REDMOND AND J.E. YUKICH, *Limit theorems and rates of convergence for subadditive Euclidean functionals*, Annals of Applied Prob., to appear (1993).

[8] W.S. RHEE, *A matching problem and subadditive Euclidean functionals*, Annals of Applied Prob., 3 (1993), pp. 794–801.

[9] J.M. STEELE, *Subadditive Euclidean functionals and nonlinear growth in geometric probability*, Annals of Prob., 9 (1981), pp. 365–376.

[10] J.M. STEELE, *Growth rates of Euclidean minimal spanning trees with power weighted edges*, Annals of Prob., 16 (1988), pp. 1767–1787.

[11] J.M. STEELE, *Euclidean semi-matchings of random samples*, Mathematical Programming, 53 (1992), pp. 127–146.